Inshore Fisheries Management

Reviews: Methods and Technologies in Fish Biology and Fisheries

VOLUME 2

Series editor:

Jennifer L. Nielsen
U.S. Geological Survey
Biological Resources Division
Anchorage, Alaska

Inshore Fisheries Management

Edited by

David Symes
Department of Geography
University of Hull, United Kingdom

and

Jeremy Phillipson
Centre for Rural Economy,
University of Newcastle, United Kingdom

KLUWER ACADEMIC PUBLISHERS
DORDRECHT / BOSTON / LONDON

A C.I.P. Catalogue record for this book is available from the Library of Congress.

ISBN 1-4020-0128-2

Published by Kluwer Academic Publishers,
P.O. Box 17, 3300 AA Dordrecht, The Netherlands.

Sold and distributed in North, Central and South America
by Kluwer Academic Publishers,
101 Philip Drive, Norwell, MA 02061, U.S.A.

In all other countries, sold and distributed
by Kluwer Academic Publishers,
P.O. Box 322, 3300 AH Dordrecht, The Netherlands.

Printed on acid-free paper

Printed in the Netherlands.

General Series Preface

Reviews: Methods and Technology in Fish Biology and Fisheries published by Kluwer Academic Publishers is a book series dedicated to the publication of information on advanced, forward-looking methodologies, technologies, or perspectives in fish and fisheries. This series is especially dedicated to relevant topics addressing global, international concern in fish and fisheries. Humans continue to challenge our environments with new technologies and technological applications. The dynamic creativity of our own species often tends to place the greatest burden on our supporting ecosystems. This is especially true for aquatic networks of creeks, lakes, rivers and ocean environments. We also frequently use our conceptual powers to balance conflicting requirements and demands on nature and continue to develop new approaches and tools to provide sustainable resources as well as conserve what we hold most dear on local and global scales. This book series will provide a window into the developing dynamic among humans, aquatic ecosystems (both freshwater and marine), and the organisms that inhabit aquatic environments.

There are many reasons to doubt the increasing social and economic value technology has gained over the last two centuries. Science and technology represent stages in human development. I agree with Ernst Mayer when he said in *Toward a New Philosophy of Biology* (1988) that "endeavors to solve all scientific problems by pure logic and refined measurements are unproductive, if not totally irrelevant." Living aquatic systems are extremely complex environments that appear alien to most human beings. We only "go there" in a limited capacity and for relatively short periods of time. To reduce these biological systems to simple physio-chemical processes that can be controlled by technology, and made subject to arbitrary management decisions or broad arm-waving policies is to deny their unique nature. That is certainly not what is meant by "*Methods and Technology*" in this series. Rather, I want to provide a forum for discussions on the living systems themselves and the organisms inhabiting them through new applications of science and technology with special emphasis on aspects of fish and fisheries under-represented in the current literature.

Our understanding of aquatic biology in freshwater and marine environments demands a careful and protracted approach ranging broadly from studies of genes, regulatory processes, isolating mechanisms, individual behavior, population structure, biodiversity, to interactive ecosystems. The tools and technologies that allow these investigations change rapidly, always removing old uncertainties and creating new ones. The intent of this series is to monitor that change and document perspectives and developments that mark a fundamental re-evaluation of nature in aquatic habitats and its role in relationship to human society and resource management.

Dr. Jennifer L. Nielsen, Series Editor
Reviews: Methods and Technology in Fish Biology and Fisheries
US Geological Survey, Biological Resource Division
Alaska Biological Science Center
Anchorage, Alaska

TABLE OF CONTENTS

PART II: COMMON THEMES: A CROSS-CULTURAL ANALYSIS OF INSHORE FISHERIES

PART III: MANAGEMENT ISSUES

PART IV: CONCLUSION

LIST OF FIGURES

LIST OF TABLES

LIST OF CONTRIBUTORS

Katia Frangoudes
OIKOS Environnement - Ressource
65, rue de Saint Brieuc, CS 84215
35042 Rennes Cedex, France

Thomas Højrup
Department of Archaeology and Ethnology
University of Copenhagen
Vandkunsten 5
1467 Copenhagen K
Denmark

Kirsten Monrad Hansen
Thorup Strandvej 267
DK - 9690 Fjerritslev
Denmark

Kjell Nybacka
Ostrobothnia Employment and Economic Development Centre (Fisheries Section)
Hovraettsesplanaden 19 A
P.O.Box 131, FIN-65101 Vasa
Finland

Jeremy Phillipson
Centre for Rural Economy
Dep't of Agricultural Economics and Food Marketing
University of Newcastle
Newcastle upon Tyne, NE1 7RU
United Kingdom

Laura Píriz
National Board of Fisheries
SUCOZOMA Programme
Box 423
40126 Göteborg
Sweden

Eva Roth
Department of Environmental and Business Economics
University of Southern Denmark
Niels Bohrs vej 9, 6700 Esbjerg
Denmark

Nathalie Steins
Dutch Fish Product Board
Postbus 72
2280 AB Rijswijk
The Netherlands

David Symes
Department of Geography
University of Hull
Hull
HU6 7RX
United Kingdom

Mireille Thom
Ceol-an-Uillt
Nethybridge
Inverness-shire
PH25 3EQ
United Kingdom

Gunnar Thoresson
National Board of Fisheries
Institute of Coastal Research
G:a Slipvagen 19
S-740 71 Oregrund
Sweden

Rob van Ginkel
Department of Anthropology
University of Amsterdam
Oudezijds Achterburgwal 185
1012 DK Amsterdam
The Netherlands

SERIES EDITOR'S PREFACE

All countries with a coastline have multiple inshore fisheries falling under direct federal or state management. These fisheries tend to be small, both in harvest and economic value, in comparison with large demersal and pelagic species harvested in offshore waters. Most of these fisheries are localised activities where fishers have limited knowledge of factors leading to distribution, abundance, and behaviour of the target species in relation to larger scale species and ecosystem issues. These fisheries are frequently small, family-owned activities or cooperatives that persist through significant changes in resource abundance and the economic value of their harvest because the fishery is rooted in community and social values that spread far beyond the effects of the fishery alone. In that sense, inshore fisheries are a social-biological phenomenon not linked to the economic avarice of larger scale fishing for aggregate resources in the open ocean. That does not mean the different aspects of inshore fisheries are of "small" concern, indeed recent literature in fisheries policy and science has held up inshore models as case studies of how traditional, community-policed fisheries can help provide sustainable harvests. Society as a whole, however, has put a low priority on research and assessment of inshore stocks, at least until that is all that is left to harvest after broad scale depletion of other, more economically viable fisheries.

The diverse mosaic of legal and political jurisdictions controlling this fishery in Europe confounds the management of inshore fisheries. There has also been a recent movement to implement a more global approach, in the European context, to fisheries policy and management. This volume represents the first thorough approach at documentation of the diversity of management structures for inshore fisheries in Europe. As such it portrays all of the deficiencies, myths, and lack of clear scientific data associated with inshore fisheries found throughout Europe. To address these deficits various authors suggest a more integrated approach that balances the social, economic and environmental needs across large ecosystem scales for the management of this important European fishery. Because of the long history of inshore harvest and the deep cultural ties to fishing found throughout Europe, the "reinvention" of inshore fisheries management in this area may well serve as a model for future fisheries in transition throughout the world.

Dr. Jennifer L. Nielsen, Series Editor
Reviews: Methods and Technology in Fish Biology and Fisheries
US Geological Survey Biological Resource Division
Alaska Biological Science Center
Anchorage, Alaska

PREFACE

This volume has its origins in two workshops held under the auspices of the European Social Science Fisheries Network (ESSFiN), a Concerted Action funded by the European Commission under the FAIR Programme. During the earlier stages of the Concerted Action frequent reference had been made to the importance of inshore fisheries in Europe but no attempt had been made to document the nature of these fisheries and their management. As a result, a small working group was convened in Gruissan in southern France in March 1998 with the purpose of comparing the nature and status of inshore fisheries and the characteristics of their management in seven west European countries. The results of this initial exercise came as something of a surprise. Not only were Europe's inshore fisheries subject, as expected, to huge diversity, but there was considerable variation in the institutional arrangements for their management, with a somewhat stronger level of central direction than many of us had anticipated and relatively weak involvement by fishermen themselves. But possibly the most surprising feature was the lack of consensus over the definition of inshore fisheries and a consequent inability in most countries to put a precise value on their contribution to the fisheries sector. The group met again in September in Amsterdam, partly to review the earlier findings but mainly to search for common denominators and to identify some key issues facing the management of inshore fisheries.

The chapters which make up the volume have evolved from presentations at the two workshops, with considerable rewriting and updating of the original drafts. The editors would like to record their gratitude for the unfailing promptness and civility with which the authors responded to requests for additional information and recommendations for changes to the text, often under what may have appeared at times to be undue provocation. We would also like to pay tribute to the contributions to the two workshops by members not represented among the authors of individual chapters - Karine Dusserre of the Projet du Parc Naturel Régional du Pays Narbonnais, Duncan McInnes of the Western Isles Fishermen's Association and Erland Eklund of the University of Helsinki - and also express our appreciation of Keith Scurr's draughtsmanship in preparing the maps and diagrams for publication. Finally, we must acknowledge the financial assistance of the

European Commission through its grant establishing the Network (FAIR CT95-0070) under the Fourth Action Programme.

David Symes
The University of Hull

Jeremy Phillipson
The University of Newcastle

INTRODUCTION

Chapter 1

INSHORE FISHERIES IN EUROPE AT THE TURN OF THE CENTURY

David Symes

1. INTRODUCTION

As far as fisheries are concerned, throughout much of the twentieth century the focus of attention in Europe has been on the offshore sector, initially as a reflection of the rapid industrialisation of the fishing industry, the growth of fishing effort and the contagious spread of overfishing, then in the 1970s as a result of the 'nationalisation' of fishing rights through the declaration of 200 mile Exclusive Economic Zones, and most recently with the realisation of a comprehensive Common Fisheries Policy (CFP) in 1983 which now embraces most of Europe's Atlantic waters. As a consequence the key issues have been those raised by the continuing failure of fisheries management - nationally and internationally - to resolve the problems of resource depletion on a global scale and, in the European context, the efficacy of legal and institutional frameworks for fisheries management created by the European Union (EU) and their interactions with Member State institutions.

In the broader global or macro-regional scheme of things, inshore fisheries have been largely ignored - dismissed as unimportant in terms of the overall volume of production, irrelevant in relation to the main thrust of the CFP, and even anachronistic in the light of structural developments in modern economies. A less dismissive portrayal would still tend to describe inshore fisheries as in denial of the realities of a modern fishing economy - a declining, traditional and somewhat introverted sector, caught up in the problems of access to scarce resources, marginalised by the processes of modernisation and economic rationalism, and struggling for political recognition. At best, inshore fisheries are portrayed as the victims of an economic sector in transition. The reality, however, may be somewhat

D. Symes and J. Phillipson (eds.), Inshore Fisheries Management, 3–23.

different. Certainly it is difficult to explain the survival of a significant and surprisingly dynamic inshore sector in the context of increasing economies of scale which have attended virtually every sector of the European economy - including fishing - throughout much of the twentieth century. Indeed, as van Ginkel (Chapter 10) and Hansen and Højrup (Chapter 11) argue, the resilience of the sector has rather more to do with the persistence of values and behaviours rooted in traditional concepts of family, community, independence and self-esteem than with the materialistic parameters of quality of life that have pervaded urbanised European societies throughout much of the last century. And it would be wrong to characterise inshore fisheries as inimical to technological change and unresponsive to the benefits of specialisation, economic efficiency and market awareness.

But in some respects inshore fisheries have been neglected. Fisheries science is largely geared up to serve the needs of formal systems of fisheries management. As a result a significant share of the scientific resources is directed towards refining our knowledge and understanding of those demersal and pelagic species which form the basis of Europe's offshore industries and, more prosaically, in the reiterative process of stock assessment. Much less time and money is spent on research and assessment of inshore stocks, especially shellfish. Meanwhile, the interests of the social sciences largely revolve around issues of renewable resource management and access to common pool resources - both of relevance to inshore fisheries but conducted mainly in the context of the offshore sector - together with analysis of international competition and collaboration and the institutional arrangements for fisheries management. Only in this last area has due attention been paid to inshore fisheries where traditional systems of management have provided case studies of co-management and self-management regimes.

In truth we know very little about our inshore fisheries. The relative dearth of scientific knowledge, deficiencies in terms of accurate assessments of their economic significance and the intermingling of myth and reality in our perceptions of inshore fisheries make for an unsatisfactory basis for their management.

All this may be about to change. On the one hand, the depletion of fish stocks, the realisation of the need to reduce fishing capacity and curb fishing effort, together with the risks of displacing fishing effort to inshore waters increases the pressure on the inshore zone and prompts questions of how to accommodate or constrain the pressure. On the other hand, it seems more than likely that the momentum for a more integrated approach to fisheries management - incorporating an ecosystem approach, seeking to balance economic, social and environmental objectives and generally realising the concept of 'responsible fisheries' - will impact first and foremost on inshore waters. In one sense, therefore, inshore fisheries and their management may well be facing a significant period of transition.

The aim of this volume is to redress the balance and initiate a better understanding of inshore fisheries in Europe through a comparative analysis. The seven countries which form the basis of the analysis are a somewhat biased sample of European experience. They are drawn exclusively from the EU: thus the coverage excludes 'independent' fishing nations like Norway and Iceland. Moreover there is a strong north west European emphasis on countries which border the North Atlantic and the tributary Baltic, North and Irish seas. Regrettably, though not intentionally, Spain - the EU's largest fishing nation - is not included, but the example of France permits a glimpse of the very different, 'other world' of Europe's fishing industry in the Mediterranean. Nonetheless, the seven countries do manage to convey something of the complexity of the circumstances in which inshore fisheries are conducted and the variety of management approaches, which have as much to do with differences in political culture and public perception as with physiographic variation and biological diversity.

2. THE NATURE OF INSHORE FISHERIES

2.1 Problems of definition

As will become strikingly evident from the national case studies (Chapters 2 - 8), one of the besetting problems of inshore fisheries management in Europe is the lack of clarity and uniformity in the definition of inshore fisheries. Several alternative criteria are in common use - vessel size, trip length, activity patterns, gears and species - and although we can use all of these to build a typical profile of inshore fisheries which may hold good for most European countries, such an amorphous delineation of inshore fisheries is of little practical value when it comes to their regulation. But the problem is not simply that different parameters are being used across Europe but also that, for each parameter, different criteria may be employed to distinguish inshore from offshore. For example, in terms of vessel length 8, 10, 12 or even 15 m may be used to define an inshore vessel. The reasons for this ambiguity are quite simple: vessels which formerly were confined to working close to the shore have, as a result of technological change and especially increased engine power, been liberated. Twelve metre 'fast boats' are capable in appropriate weather conditions of working up to 40 miles offshore. Trips of less than 24 hours duration - another common element in the conventional definition of inshore fishing - can by the same token extend well beyond what are normally considered to be inshore waters. Similarly, species which used to be taken almost exclusively in coastal waters by small boats deploying simple gears are now being targeted well offshore: the European lobster is a case in point.

The traditional generic definition of inshore fishing, commonly based on vessel length, is therefore no longer appropriate. At best it is a crude approximation to reality; at worst it can harbour potentially dangerous misrepresentations concerning the nature of inshore fisheries and their environmental impacts. Unfortunately generic definitions are still widely used in European and national fisheries legislation. In their place we need to rely much more on a geographical definition referring to the legal territorial limits of inshore waters, though this too is subject to some inconsistency across Europe. As a consequence inshore fisheries management would become more firmly grounded in a zonal rather than structural approach. However, apart from certain socio-economic and ecological benefits attached to a distinctive management regime for inshore waters, the separation of inshore and offshore fisheries is to some degree artificial, for the biological resources of the inshore zone form part of a continuum of resource interdependence and interaction which stretches from the neritic zone to the continental slopes. Globally it is estimated that around two thirds of all marine fish begin their life cycles in shallow coastal waters and estuaries.

2.2 The characteristics of inshore fisheries: romanticism and reality

Different terminologies are used interchangeably to describe inshore fisheries: 'inshore', 'small boat', 'small scale', 'artisanal'. Each conveys a particular element of the overall situation and, taken in combination, they offer a reasonably sound and comprehensive description, but one which can contain overtones of romanticism. As the most immediately visible and easily visualised sector of the industry, inshore fisheries tend to be encapsulated in the popular 'picture postcard' imagery of fishing based on childhood memories of seaside holidays and the fascination of harbours with often brightly coloured, intriguingly named small fishing boats - an image still capable of exerting a powerful tourism attraction. It is certainly true that the inshore sector provides a rich reservoir of folkways with considerable regional variety among traditional craft and gears - in marked contrast to offshore vessels which over time have evolved into more standardised forms. Even among the smaller craft, this regional variety is being eroded with the new generation of fibreglass fast boats built to a more or less standard design.

But it is not so much their physical appearance but the behavioural characteristics that distinguish Europe's inshore fisheries. Two features above all others typify inshore fisheries: the polyvalent nature of the inshore fleet and the essential flexibility of the sector. All manner of fishing gears are deployed to harvest a variety of species at varying depths in the water column: the same boat may engage in strikingly different types of fishing activity, usually in different seasons, but even on occasions during the same day. Passive gears (fixed nets, lines and traps) and encircling gears (seines and *lamparos*) predominate. Mobile gears may prove difficult to operate in inshore waters with their complex currents and tides; or they may require larger, more expensive boats; or indeed they may be outlawed by local management rules designed to protect the static gear operators.

The flexibility and dynamism of the inshore sector is at once both expected and rather surprising - expected in the sense that it arises naturally from the polyvalent nature of inshore fishing and from the need to combine several different fishing seasons in order to create the basis for a year round occupation, and surprising because traditional, small scale economic activities are often noted more for their tendency towards retrenchment than innovation in response to difficult conditions. The flexibility of inshore fisheries may also reflect a continuing or former involvement in pluriactive coastal economies in which fishing is combined with farming or other activities, often on a seasonal basis. Moreover, inshore fisheries formerly provided a buffer economy to be exploited more intensively or by greater numbers when other sectors of the regional economy were doing less well and were unable to absorb the surplus rural labour. Although pluriactivity has generally declined across Europe, it can still exert a strong influence on peripheral coastal economies. In many European countries there is a much higher incidence of part time, seasonal or casual participation in inshore fishing than in almost any other sector of the economy. In part this explains the difficulties in providing accurate statistical assessments of the inshore sector. It is also a source of perplexity and irritation for fisheries managers, leading them to 'write off' part time fishermen as 'non-professionals' and to ignore them when making management decisions.

Inshore fishermen have demonstrated a remarkable aptitude for change, exploiting new species, opening up new market niches and in general maintaining a vital role in the supply of high quality fresh fish to local and national markets. They have adapted many of the technological innovations initially intended for the offshore sector thus enabling them to extend their operational range and to intensify their exploitation of inshore waters. In some senses, however, this very flexibility can become a problem for inshore management. Even with licensing schemes restricting entry to the sector, the relative ease with which inshore fishing units can diversify into alternative species or intensify their fishing effort through new technology, greater engine capacity or simply increasing the number of hours fishing makes inshore fisheries somewhat unstable. There are clear dangers that this capacity for flexible response can all too easily translate into overcapitalisation of the inshore sector. Nonetheless, flexibility is still to be regarded as its cardinal strength. Resilience - in face of changing natural environmental conditions, market fluctuations, growing competition and encroachment by offshore interests and increasing regulation - forms the basis for the sector's survival.

2.3 Socio-economic assessments of inshore fisheries: the problems of inadequate data

Despite these apparent behavioural strengths - or possibly because of them - evaluation of the inshore sector in economic and social terms is difficult to achieve. This may in turn lead to profound misjudgements of the significance and role of inshore fisheries in terms of their overall contribution to the fishing industry and to the regional economies. It also calls into question the seriousness with which governments treat the sector. Explanations for the lack of reliable statistical data on inshore fishing fleets, employment, catches and fishing effort, which must surely pose serious problems for management, reflect many of the circumstances discussed above, including the absence of a standardised definition of inshore fisheries, the fragmentation of the sector and the uncertain boundaries between commercial, professional fisheries on the one hand and subsistence and recreational fisheries on the other. The fragmentation of the sector represents a major obstacle: the large numbers of boats making irregular deliveries of small, diversified catches to widely dispersed landing points makes the collection of data difficult. We also need to bear in mind that a significant part of the total catch from inshore boats is handled through informal, grey markets involving direct transactions with retailers, restaurants and consumers where documentation in the form of sales notices is missing.

To some extent this apparent lack of professionalism may be seen as the residual of an informal, pluralistic, pre-industrial economy. It is, however, a serious aberration at a time when the fishing industry is subject to increasing regulation, monitoring and control and when fisheries management is putting itself more and more in the hands of the applied biological sciences. When it comes to inshore fisheries we have a far from perfect knowledge as to how much is being taken out of the stocks through fishing. It is perhaps not all that surprising that an activity that cannot readily be reduced to statistical norms or certainties will tend to resist attempts to formalise procedures for purposes of closer regulation.

The paucity of data on inshore fisheries raises questions of the 'chicken and egg' variety. Does the difficulty in recruiting sound statistical information contribute directly to the apparent lack of political interest in inshore fisheries and their management in today's science driven policy world? Or does the low priority granted to inshore fisheries explain the lack of determination to collect or simply collate adequate data?

Whatever the answer, there is a tendency to undervalue inshore fisheries. In part this may reflect an earlier emphasis on quantity rather than quality when evaluating fisheries and the current preference for measuring economic efficiency in terms of returns on capital. On both counts, inshore fisheries will emerge rather badly. When translated into value, the inshore sector in Europe probably accounts for no more than 25% of current production, though this of course comes from only a tiny fraction of the total surface area of the European seas. The real importance of inshore fisheries is to be measured not in terms of output but in employment and in their particular contributions to local economies especially in the remoter, more rural fishing dependent regions.

Despite the decline in the number of jobs in the inshore sector noted in nearly all the national case studies, which make up the first part of this book, overall the sector is maintaining its relative share of fishing related employment. The contribution of the local inshore fishing fleet to employment and wealth creation in the local economy, especially in areas where alternative employment opportunities are scarce, remains important. At a time when the inter-dependence between local fleets, local fishing grounds and local ports has been eroded by the rise of a much more mobile offshore sector there is a very real sense in which the inshore sector remains faithful to the local economy. The surviving linkages between inshore fisheries and the local economy are largely based on pragmatism: catches are usually too small to warrant being landed far from the local fishing grounds - though even this relationship is being threatened by the emergence of nomadic inshore fleets.

3. THE COMPLEXITIES OF INSHORE FISHERIES MANAGEMENT

Inshore fisheries management predates systems of offshore regulation by several centuries and in the Mediterranean world by two or three millennia (Collet, 1999). It is also very much more complex in form and content with strikingly different institutional approaches - centralised, decentralised and devolved - deployed across Europe (see Chapters 2 - 8). Originally inshore management was a largely local initiative, territorially based and reliant upon maintaining self-discipline among fishermen and upon the support of community institutions. It was rarely concerned directly with issues of resource conservation - there was, in fact, little need - but rather with guaranteeing equity in access to the local resources. Throughout the twentieth century, however, the state has progressively usurped the functions of local self-management institutions, partly in response to the breakdown in local authority when faced with the increasing mobility of fishing fleets and the frequent incursions of 'outsiders' into traditionally defined local fishing territories. At the same time, the emphasis of management has been reversed with modern systems focusing mainly on resource conservation and leaving aside concerns for local distributive justice.

According to Field (1989), the governance of common pool resources involves collective actions to regulate the behaviour of those who use the resource and to prevent outsiders, or those not qualified to use the resource, from gaining access to it. The balance between these two mechanisms - regulation and exclusion - is to some extent a function of social costs. When resolving conflicts between different types of user through regulation becomes difficult and transaction costs rise, the management system will tend towards greater use of exclusion as a means of controlling the pressure on resources and of separating non-compatible uses. Conversely, when users are able to regulate themselves more efficiently and so reduce the transaction costs through collective decision making and enforcement, the need to exclude certain users will decline (van der Schans, 1999). In practice, however, a prerequisite for effective and efficient regulation of fishing activities may well be the exclusion of those for whom the local resources only have a transitory importance and who are therefore unwilling to acknowledge the legitimacy of the local management institution. This was in fact a basis for the success of traditional management systems such as those developed by the *prud'homie* in Mediterranean France (Chapter 8). Modern inshore management tends to rely more on regulation largely because it has proved difficult in the past to resist the claims for access made by an ever widening range of users of inshore waters, though the principle of exclusion may still be exercised in the zonation of uses.

3.1 Complexities of inshore fisheries management: (i) internal diversity

The complexity of inshore management derives from two quite different circumstances: the internal diversity of inshore fisheries and external pressures induced by the growing range of legitimate uses for inshore waters. These inshore waters contain a far greater variety of fisheries, each with its own particular biological characteristics, than is usually associated with offshore waters. They include finfish fisheries for the mobile demersal and pelagic species and for highly migratory species like salmon and eel; shellfish fisheries for the sessile crustacean species (crab, crayfish and lobsters) and for the molluscan species (mussels, oysters and cockles) which are more or less static at least during their post-larval phases. Each of these requires a different form of management which reflects the geographical distribution of the unit stock, its basic biological parameters and the variety of fishing methods involved. The problem is further complicated by the fact that these different fisheries are not neatly segregated into separate segments of the inshore fleet but usually combined in the same polyvalent enterprise. Moreover, the management strategies must also take account of the fact that inshore waters often contain the spawning and nursery grounds for species which form integral parts of the offshore fisheries.

To add still further to the complexity of inshore fisheries management, responsibility for these different types of fishery may not necessarily be entrusted to the same agency: in England and Wales, for example, the highly migratory salmonid fisheries are the responsibility of the Environment Agency, quite reasonably reflecting their presence in inland and marine waters, rather than coming within the compass of the Sea Fisheries Committees (see Chapter 6). This fragmentation of responsibility can cause a number of anomalies which reduce the overall effectiveness and efficiency of the inshore management regime.

One of these anomalies concerns aquaculture, a term which embraces both shellfish production - a continuation of age old forms of mussel and oyster cultivation and which readily forms an integral part of inshore fisheries - and finfish production, most notably in western Europe of salmonids but now being extended to a much wider range of species including cod and halibut. It is the latter, much more recent development that poses particular problems for inshore fisheries (see Chapter 13). Not only has the spectacular rise of salmon aquaculture in countries like Norway, Scotland and Ireland had a dramatic effect on markets but in almost every respect the systems of production contrast sharply with our earlier images of inshore fisheries in terms of scale, structure, investment, risk, flexibility and so on.

The cultivation of fish on a fixed site is a fundamentally different kind of activity which some argue is much closer to agriculture than to fishing (Bailey *et al.*, 1996) but in some respects appears more akin to industrial production, certainly in terms of its economic and social organisation. Salmon aquaculture, in particular, with its dependence on external sources of investment, capitalist dynamic and wage earning mode of production, is a world away from the family based, share system embodied in a simple commodity production system which characterises at least some parts of the inshore fisheries (see Chapter 11). Although aquaculture is sometimes posited as a 'replacement economy' for a declining inshore sector, the overlap in terms of knowledge, skills and work practice is relatively small. According to Dwire (1996), aquaculture is commonly perceived as a major force for social change in a way that developments in capture fisheries rarely are. It touches directly on the lives of the local inhabitants and threatens to intrude upon their local value systems and quality of life.

A fundamental distinction between aquaculture and capture fisheries, and one which has proved a frequent source of conflict, relates to property rights. While for capture fisheries, the fishing grounds and the fish stocks remain to a greater or lesser extent an open access resource with ownership of the fish only resolved through catching, in modern aquaculture systems both the fish stocks and the production sites are privately owned. Privatisation and the exclusion of inshore fishermen from their traditional grounds are obvious sources of aggravation to inshore fishermen and a basic reason for their antipathy towards aquaculture development; so too are the much publicised risks of environmental degradation arising from high intensities of fish farming.

In mediating the potential conflicts, the institutional location of aquaculture in terms of departmental responsibility can prove significant (Bailey *et al.*, 1996). Not only will it determine the legislative framework but also the particular form of decision making in relation to licensing, including the extent of public participation; moreover, the outcomes may well reflect the extent of leverage which the particular department is able to apply in favour of aquaculture development. Responsibility for finfish aquaculture is often separated from overall responsibility for fisheries and vested in agriculture departments and/or devolved to agencies responsible for water resource management. It is quite usual for aquaculture to be presented as an integral part of the mechanisms for regional development in under-performing peripheral coastal areas. Governments will therefore stress the positive role of aquaculture in increasing regional output, generating export earnings and creating employment and, as a result, will tend to impose only minimum environmental standards for individual enterprises, setting aside issues of overall environmental impact and social justice.

3.2 Complexities of inshore fisheries management: (ii) competing claims on inshore waters

Not only must inshore fisheries management cope with the problems posed by the co-existence of different types of fishing activity occurring simultaneously on the same fishing grounds but it also has to contend with competing claims on marine space made by increasingly powerful interests outside the realm of fisheries. This last theme is extensively considered in Part III and there is little need therefore to rehearse the detailed arguments. Suffice it to point out that, from a position not all that long ago of a more or less unchallenged monopoly of use rights in inshore waters, fishing has become simply one among many different legitimate uses. Its hegemony is thus seriously undermined by the popular support which several of the new uses can command, as in the case of recreational fisheries and marine wildlife conservation, or by claims of higher levels of contribution to the regional economy, as with tourism. Another threat comes as a result of the evident weaknesses in the political organisation of inshore fishing interests. In terms of political organisation, inshore fisheries are neatly impaled on the horns of a dilemma. Although they sense that the national federations may not adequately represent the interests of small scale, artisanal, 'non-professional' fishermen in the key policy debates, they also recognise that forming breakaway organisations specifically designed to reflect their particular interests may run the risk of actually reducing their influence. Such would be the case where the central administration deliberately chooses to limit the consultation process to 'recognised organisations'. But as Chapter 16 demonstrates, the failure to engage in public debate and to articulate their views can lead to management decisions which fail to prioritise the inshore fishermen's most urgent needs.

Inshore fisheries management cannot be completely divorced from the management of offshore fisheries. Nor can it claim special privileges in relation to the planned use of coastal waters. Today the management of inshore fisheries is enmeshed in a matrix of management institutions arranged both vertically, in terms of geographical scale ranging from European through national to local levels, and horizontally across a number of discrete and competing sectoral interests, many of which exert greater political influence than do fisheries. The challenge for the future would therefore seem to lie in finding a formula for integrated management which permits inshore fisheries to sustain their roles in providing local employment and in supplying markets with good quality fresh fish, while coexisting with other legitimate users of coastal waters in ways that do not put the environmental quality of those waters and the diversity of their ecosystems at risk. We return to this challenge in the final chapter.

4. SHAPING THE ARGUMENT

In seeking to elaborate on some of the foregoing themes and so develop the analysis of Europe's inshore fisheries and their management, the remainder of the volume is divided into three main sections. The first describes the situation in seven European countries - all Member States of the EU (see Figure 1.1); the second attempts a cross-cultural analysis of inshore fisheries from the perspectives of social organisation, cultural patterns, economic motivation and institutional design; and the third extends the analysis of some of the key issues confronting inshore fisheries management at the start of the new millennium.

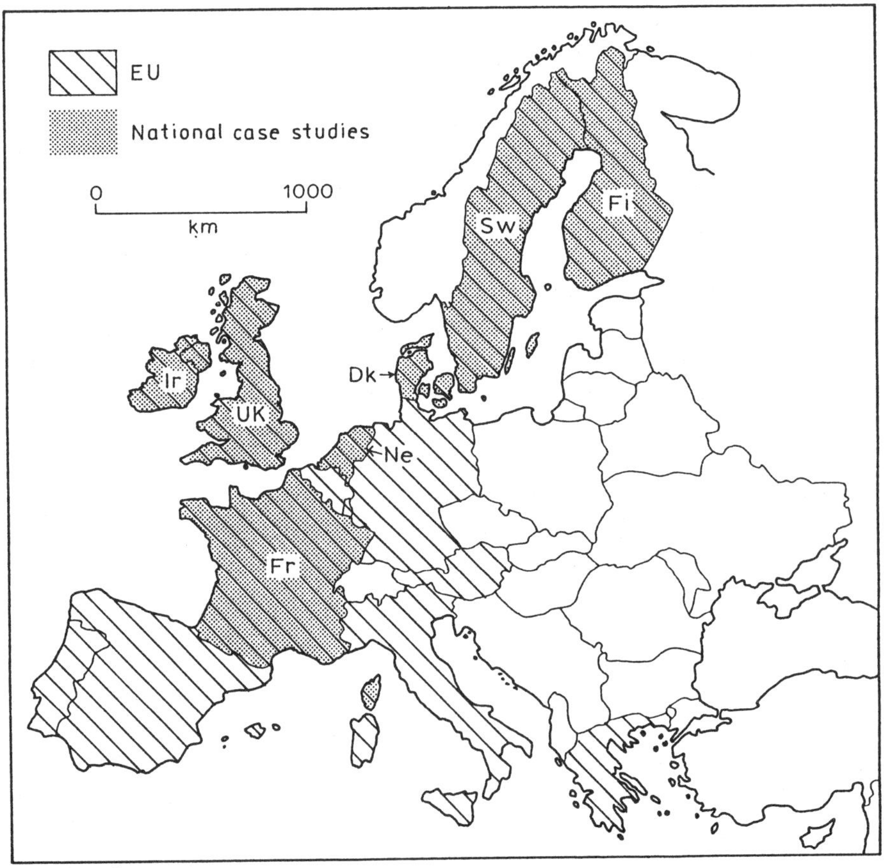

Figure 1.1 The National Studies

4.1 National studies

Although written to a broadly similar template, each chapter in Part II attempts to portray the defining characteristics of the country's inshore fisheries and their systems of management. Direct comparisons between countries may sometimes prove difficult - a situation made all the more frustrating by the different conceptualisations of inshore fisheries and the lack of standardised parameters for defining inshore fisheries in the national statistics. Nonetheless, what emerges is something of the quintessential variety of Europe's inshore fisheries, in part conditioned by natural geographical characteristics but also by the salient features of particular political cultures which govern the approach to fisheries management in each of the seven countries. Notwithstanding this variety of circumstance, a number of common denominators can be identified, together with a shared set of management issues that need to be addressed.

The first two chapters describe the situations in the two most northerly - and, coincidentally, the two most recent - Member States of Finland (Chapter 2) and Sweden (Chapter 3) where inshore fisheries are closely circumscribed by the severe conditions of the northern winter and by the particular hydrographic characteristics of the Baltic Sea, resulting in highly seasonal patterns of activity and a relatively narrow range of target species. Apart from the physiographic constraints, two other features serve to distinguish the nature of inshore fisheries in Finland and Sweden: first, the distinctive pattern of private ownership of waters closest to the shore; and, secondly, the considerable influence exerted by recreational fisheries in countries where the ownership of summer homes, preferably on the lake or sea shore, is a key element in the quality of life. Here fishing is considered not just part of the recreational experience of second home ownership but also as contributing to household consumption patterns. Private ownership and recreational fishing are both seen as adding greatly to the problems of inshore management. However, the experiences of the two countries diverge. In Sweden, individual owners retain responsibility for access and use of private waters, while in public waters the inshore fisheries are largely subject to central direction. By contrast, management of public inshore waters in Finland is devolved to Fisheries Districts and regulation of private waters is nominally in the hands of statutory fishing associations, though these play only a minor role in the overall management system.

Although by European standards Denmark (Chapter 4) is a relatively small country, its intricate geographical structure of peninsulas and islands and its location between the North and Baltic Seas help to establish its credentials as one of Europe's foremost fishing nations. Although no formal distinction is drawn between inshore and offshore fisheries in the basic management system - which remains characterised by a centralised top-down model - paradoxically the account is one of the very few which is able to describe in detail the contribution of the inshore sector to the overall pattern of landings. Once again, the need in the management regime to mediate between the interests of commercial and other fishing interests forms an important theme. Centralised management does not appear to detract from the sensitivity to local conditions, as exemplified by the abundance of local regulations and the recently completed integrated management plan for Limfjord in an attempt to balance the competing claims of nature conservation, blue mussel production and recreational use of the area.

In the case of the Netherlands (Chapter 5) the comparatively short North Sea coast is interrupted by estuaries, fringed by islands and greatly modified by large scale engineering projects for land reclamation and flood defence: altogether, a quite extensive area falls within the definition of inshore waters. By contrast with all other countries included in this volume, the Netherlands has a remarkably compact and highly rationalised fishing fleet both in terms of its structure and the distribution of fishing rights in the form of individual transferable quotas for the most important finfish species. Dutch inshore fisheries are synonymous with shellfish, especially mussels and cockles. A unique system of integrated management has been developed in the 1990s in which inshore fisheries management is closely correlated with marine wildlife conservation objectives, notably in the Wadden Sea, and where specialised producers' organisations have been encouraged to develop their own management plans for the mussel and cockle fisheries. In other words, the inshore fisheries are largely self-regulated, though overseen by central government agencies; but harmonisation of fishing and wildlife conservation interests is still incomplete.

Inshore fisheries management in Britain (Chapter 6) is complicated by the division of responsibility between the four constituent countries - England, Wales, Scotland and Northern Ireland - which predates the process of political devolution in the 1990s. Overall, the inshore fisheries in general and the shellfish sector in particular have exerted a stabilising influence during a prolonged period of decline in offshore fishing opportunities. The analysis focuses mainly on England and Wales where management of inshore fisheries has been decentralised and devolved to Sea Fisheries Committees (SFCs) for well over a century. The result is a form of bureaucratic co-management where local authority representation is balanced by representatives of fishing (and other) interests nominated by the central administration rather than chosen by the industry's own organisations. By contrast in Scotland, inshore fisheries management has until very recently remained faithful to the principle of central direction with little formal consultation with the industry; growing pressure for locally or regionally devolved management through shellfish Regulating Orders is unlikely to involve creating statutory bodies similar in style to SFCs south of the border.

Ireland (Chapter 7) with its long, irregular coastline had until quite recently largely turned its back on the sea as a resource base for its developing economy. During the final decades of the twentieth century, however, the fishing industry has begun to share in Ireland's reputation as the 'tiger economy' of the EU with a rapid expansion of both the offshore and inshore sectors, coupled with a significant growth in aquaculture, which has meant a tripling of the volume of production in just 25 years. Important catalysts for the local development of inshore fisheries have been the fishing cooperatives especially active in the production and marketing of shellfish. To some extent, however, the modernisation of Ireland's fishing industry has outstripped developments in the institutional infrastructures for management, especially in the inshore sector. The need for a well defined strategy for inshore fisheries has been recognised with recent proposals to establish an inshore fisheries committee to advise the Minister and, at the local scale, the piloting of three voluntary inshore development committees.

The inclusion of France (Chapter 8) provides a tantalising glimpse of contrasts between the inshore fisheries of the Atlantic and Mediterranean coasts of Europe. In common with all other countries, inshore fisheries in France suffer the disadvantage of confused, and at times conflicting, definitions based on technical and geographical parameters. Regional analysis indicates that the polyvalent small boat sector plays a relatively greater role on the Mediterranean coast. Important variations in the management of Atlantic and Mediterranean fisheries reflect basic differences in natural conditions. In the Mediterranean region, the truncated continental shelf and the absence of 200 mile Exclusive Economic Zones, together with the 'incomplete' development of the CFP, tend to place greater responsibility for management on the national and regional administrations. Two particular features distinguish fisheries management in France: first, its strongly hierarchical structure reflected in both governmental and professional organisations; and, secondly, the significance of syndicalist forms of organisation. Here the strongest influence is exerted by the *comités des pêches* which bring professional interests to bear on management at several levels and, in the Mediterranean region, the *prud'homie*, among Europe's oldest management institutions. The latter provide, in effect, a system of territorially based self-management especially in the brackish water *étangs* which fringe much of France's Mediterranean coast.

Three broad areas of concern for the future of inshore fisheries can be identified from the national accounts - some of these are dealt with more fully in later sections of the book. First, inshore fisheries are universally recognised as a declining area of economic activity, judged in terms of participation levels; as a result the age structures of inshore fishermen are becoming significantly older and there is an urgent need for the renewal of capital investment particularly in new boats and improved quayside facilities in order to encourage new entrants to the sector. Secondly, there is a heightened awareness of the potential collision of interests between inshore fisheries and environmental conservation: as the most visible activity in inshore waters, the fishing industry has become a soft target for conservationists. Although there is growing evidence of efforts to integrate inshore fisheries management with concerns for marine wildlife conservation, there are also disturbing signs that even in the most ambitious of these projects consensual arrangements may be insufficient to placate the demands of some environmental organisations. Thirdly, there are the consequences of accelerating rates of urbanisation of Europe's coastline which manifest themselves in different forms but principally in the increasing pollution of coastal waters and in the impacts of tourism and recreational pressures on marine and foreshore space and, in the case of recreational fisheries, competition for resources.

Somewhat surprisingly there is little apparent concern for the competition between inshore and offshore sectors for access to resources or markets. Rather, the sense of competition is contained within the inshore sector in the conflicts between traditional capture fisheries and modern forms of aquaculture development. Nor is there any strong iteration of the argument that modern systems of fisheries management tend to suppress the essential diversity and flexibility of small boat fisheries and thereby marginalise the inshore sector. On the other hand, there is an inference running through most of the accounts of the need to develop distinctive policies for inshore fisheries and for their incorporation within a broader concept of integrated coastal zone management.

4.2 Cross-cultural analysis

In Part III, cross-cultural analyses of inshore fisheries tend to stress their embeddedness in specific social, cultural and economic contexts and their responses to the pressures for modernisation. This is particularly the case in Chapter 9 where the processes of social reproduction of inshore fisheries and the forms of social organisation are reviewed by Symes and Frangoudes. Family and community are seen to provide the essential building blocks of social organisation with boat crews ideally constructed around a nucleus of kin drawn from the family household or extended family network. Today the established systems of social organisation and reproduction are threatened internally by widespread changes affecting the structure of the domestic household, the changing occupational roles of women and the dilution and dispersion of kinship ties. Externally they are challenged by resource depletion and the usurpation of local regulation by state led management institutions. Crew sizes have fallen as a reflection of problems in covering the costs of additional crew members and also in some areas because of a much broader range of occupational choice than previously. As a consequence new survival strategies are required to confront the declining economic and social prospects of inshore fishing.

In Chapter 10, van Ginkel continues the analysis of traditional inshore fisheries as signifying not merely a job but also a way of life in which fishers derive satisfaction from their work and their identity as fishers, with its connotations of independence, self-reliance and freedom which may in part compensate for low financial rewards. Occupational communities of fishers are often characterised by an egalitarian ideology which emphasises equivalence and plays down status differences between crews. Despite the individualism commonly ascribed to fishers, inshore fishermen do cooperate and can be regarded as 'cooperating competitors' The interplay of competition and cooperation holds true for the access to and use of resources and for behaviour in the market place. Individual competitiveness may, however, frustrate attempts to formalise cooperation and also render inshore fishers suspicious and resentful of bureaucratic intervention in the form of management. With the social changes described in Chapter 9, what was formerly a dominant local culture has become an occupational sub-culture; yet despite the decline in inshore fishing, many fishers tenaciously cling to fishing as a job - for them fishing is a lifestyle and an enduring part of their self-image.

One of the problems experienced by modern management approaches to inshore fisheries is the apparent difficulty of their incorporation within conventional assessments of economic performance based primarily on returns on capital. In Chapter 11 Hansen and Højrup offer an explanation for the persistence of inshore fishing often in defiance of economic rationality, through life mode analysis. They portray inshore fisheries as simple commodity production in which the skipper-owner may undertake a range of functions (marketing, repair and maintenance) normally hived off to third parties. It is a form of production relatively impervious to market fluctuations and able to function for long periods without earning revenues commensurate with the value of the capital investment. The strength of simple commodity production lies in its intimate association with family enterprise. Confronted by falling prices producers have a range of options open to them - either to increase output or subsidise the fishing enterprise through undertaking other activities, including wage work. Simple commodity production is thus able to exploit its particular socio-cultural advantages through flexible organisation.

Finally in this section, Phillipson and Thom (Chapter 12) re-examine the evidence contained in the national studies in order to construct a typology of regulatory regimes occurring in Europe's inshore waters. The basic choice between central and local approaches to management is by no means an obvious one, with strong arguments in favour of either approach. Indeed, in few cases is the resulting system a simple case of either centralised decision making or delegated local authority. In practice all systems are subject to some degree of central direction and few are devoid of any element of local influence. Significantly, most of the institutional structures have been devised with a rather narrow, traditional view of fisheries management in mind. Interest in the future will revolve around the question of what kind of management framework will be best suited to taking on board new areas of management responsibility including, for example, the ecosystem approach.

4.3 Management issues

As indicated earlier the national studies uncovered a number of shared concerns for the future development of inshore fisheries in Europe. Most of these refer in some way to external pressures and to competing demands for the use of inshore waters. Certain of these issues - the competition from aquaculture, the potential threats from an ecosystem approach to inshore fisheries management and the stream of conflicts arising from the multiple use of inshore waters - are explored in subsequent chapters. Taken together these issues point to a common, though as yet elusive, solution in the form of integrated coastal zone management. Within Europe the country which has probably gone furthest down the road to achieving a workable system of coastal management is Sweden; Sweden's experience in this field is subject to a critical assessment in the final chapter.

The dramatic growth of aquaculture production in the last two decades of the twentieth century and its implications for established forms of inshore fisheries are examined by Steins and Roth (Chapter 13). To date in Europe, aquaculture has focused almost exclusively on the cultivation of high value shellfish - especially mussels, oysters and clams - and on salmonid finfish. It is the latter that has had the most far reaching consequences and, with the extension of the production techniques to a wider range of species - including cod, haddock and halibut - is likely to pose even more searching questions for the future of inshore fisheries. The authors review the benefits and disbenefits of marine aquaculture in respect of the generation of employment and income, impacts on markets and market prices, and environmental externalities *inter alia*. More particularly, they focus on the extent to which modern aquaculture can be accommodated as a complementary activity in the existing economic, social and cultural contexts of inshore fisheries. In terms of management responsibility and practice, however, aquaculture and inshore fisheries are quite distinct, usually coming under the aegis of different departments and subject to very different forms of regulation.

The environmental externalities arising from aquaculture and capture fisheries and the need for a closer integration of fisheries management and marine wildlife conservation are currently being addressed through the development of an ecosystem approach to fisheries management, here reviewed by Symes in respect of inshore waters (Chapter 14). Inshore fisheries in Europe are conducted in waters containing some of the most highly diverse, sensitive and potentially vulnerable marine ecosystems which are now the focus of growing concern for the protection of their productivity, diversity and integrity. The potential risks from fishing activity are well documented but the evidence of political action to reduce or exclude these risks is unconvincing. On the wildlife front important steps have been taken to develop a network of protected sites under the EU's Natura 2000 programme and to draw up voluntary Biodiversity Action Plans for the conservation of threatened marine fauna and flora. But there is a pressing need to ensure the success of these initiatives and secure the long term sustainability of the marine ecosystems through the development of operational objectives for an ecosystem approach to the management of Europe's inshore waters.

Interactions between fishing and the marine ecosystem are just one aspect of the need to develop a more integrated approach to the management of several competing activities in inshore waters which, if not carefully regulated, will result in environmental degradation and loss of economic efficiency. In Chapter 15 van Ginkel and Steins place fisheries at the centre of the web of potentially damaging interactions and they systematically examine the impacts sustained as a result of internal conflicts; the development of aquaculture; the extraction of non-renewable resources; non-extractive uses including tourism, leisure, residential, industrial and transport developments; and the abuse of the marine environment through waste disposal and pollution. Apart from the last of these, all are to a greater or lesser extent legitimate and necessary uses of inshore waters and ways must therefore be found for effective multi-use management through inter-sectoral collaboration. Successful integrated management, it is argued, requires the creation of decentralised and open structures.

Much is made of the need to resolve the increasing pressures on marine space through integrated coastal zone management. However, Píriz's analysis of the Swedish coastal management system (Chapter 16) suggests that inshore fishers' interests are not always adequately reflected in the outcomes of the planning process. Her chapter examines how the legal instruments and procedures for coastal management impact on - or, in some cases, ignore - the specific needs of inshore fisheries. The system of coastal planning is built around the landmark Natural Resources Act, which takes a utilitarian rather than ecocentric approach, and implemented through comprehensive municipal plans still based on the principles of land use planning. While the intention may be to encourage public participation in the planning process, in effect coastal planning becomes a closed bureaucratic exercise due largely to poor attendance at public meetings, especially by local fishing interests. The self-exclusion of fishers from the public policy process is a reflection of several factors - weak organisation, distrust of bureaucracy and an overriding sense that the ascendant issues in coastal zone management are not the same as those that exercise the inshore fishing industry.

By way of conclusion (Chapter 17) the editors return to the question of whether recent trends point to increasing convergence or divergence between the inshore and offshore sectors and to the problems posed by the growing complexities surrounding inshore fisheries and their management. For management purposes they propose a clear separation of the two sectors with a zonal system for inshore fisheries based on the primacy of social objectives in the form of employment and income generation and characterised by a combination of individual and communal use rights reserved essentially for local small boat fishermen and subject to local regulation built around technical conservation measures. This would contrast with a centrally managed offshore sector where the privatisation of use rights, through the adoption of individual transferable quotas, could be introduced in pursuit of economic rationalisation. However, caution is counselled in relation to overhasty attempts to incorporate several new concepts - the ecosystem approach and integrated coastal zone management included - into new institutional arrangements for inshore fisheries management at a time when the sector is deeply suspicious of the motives and sceptical of the outcomes.

5. REFERENCES

Bailey, C., Jentoft, S. and Sinclair, P. (1996) Social science contributions to aquacultural development, in C. Bailey, S. Jentoft and P. Sinclair (eds.), *Aquacultural Development: Social Dimensions of an Emerging Industry*, Westview Press, Boulder, Colorado, pp. 3-20.

Collet, S. (1999) Regionalisation and eco-development: which pathway for artisanal fisheries? in D. Symes (ed.), *Europe's Southern Waters: Management Issues and Practice*, Blackwell Science, Oxford, pp. 42-52.

Dwire, A. (1996) Paradise under siege: a case study of aquacultural development in Nova Scotia, in C. Bailey, S. Jentoft and P. Sinclair (eds.), *Aquacultural Development: Social Dimensions of an Emerging Industry*, Westview Press, Boulder, Colorado, pp. 93-110.

Field, B.C. (1989) The evolution of property rights, *KYKLOS*, **42** (3), 319-346.

Van der Schans, J.W. (1999) Governing aquaculture: dynamics and diversity in introducing salmon farming in Scotland, in J. Kooiman, M. van Vliet, and S. Jentoft (eds.), *Creative Governance: Opportunities for Fisheries in Europe*, Ashgate, Aldershot, pp. 95-118.

PART I: NATIONAL STUDIES

Chapter 2

FINLAND

Kjell Nybacka

1. INTRODUCTION

For a country with over 1,000 km of mainland coastline, a myriad of small islands and some 190,000 lakes over 500 m^2 in area, Finland's annual commercial catch of less than 110,000 tonnes is surprisingly small and its overall contribution to GNP (0.2%) meagre. Part of the explanation lies in the country's northerly aspect and the fact that most of her fishing waters are icebound for two months in the south and up to five or six months in the north. Another relevant factor is the structure of Finland's fishing fleet, dominated as it is by small multi-purpose boats primarily suited to coastal fishing: these account for almost 88% of Finland's registered fleet of around 4,000 fishing vessels.

Production statistics, however, fail to give an accurate picture of the industry's importance. Although the number of full time fishermen may be quite small, commercial fishing provides a primary or supplementary income for some 4,000 people and makes an important contribution to the regional economies of the west coast and the Åland Islands in particular. Moreover, around 400,000 Finns make regular use of summer cottages located on lake, river or seashore where fishing is likely to form part of the recreational experience and contribute to household consumption patterns. It is estimated that around two million Finns - 40% of the national population - catch fish at least occasionally (Salmi *et al.*, 2000). Thus recreational fisheries - a term which in Finland covers both traditional activity for home consumption and the more recent forms of sports fishing - play a significant role in the everyday life of many Finnish citizens and substantial catches are taken by recreational fishermen. In 1996, for example, the total recreational catch reached 61,300 tonnes of which 25% was caught in marine waters (see Fig.

D. Symes and J. Phillipson (eds.), Inshore Fisheries Management, 27–42.

2.1). This compares with an annual catch of around 110,000 tonnes landed by the so-called professional fishermen who rely on fishing for a significant share of their annual incomes. For them the principal species are the Baltic herring, cod, salmon, sprat and whitefish. The distinction between more or less full time 'professional' or part time and casual 'recreational' fishermen is an important one in Finland: it helps to structure management approaches and it fosters a sometimes acrimonious debate.

In 1995 Finland became a member of the EU and a structural programme for the fisheries sector (1995-1999) was initiated by the EU in July of that year. Although Finland has adopted the general principles and the detailed rules of the Common Fisheries Policy, transitional arrangements were agreed in relation to access to European waters which will remain in place until the implementation of the fishing permits scheme becomes effective or until the end of 2002 at the latest. Measures for the conservation and management of fishery resources in the Baltic are formulated within the International Baltic Sea Fisheries Commission (IBSFC) to which the EU is a party. Indeed the EU now represents the four Baltic Member States - Denmark, Germany, Sweden and Finland - in deliberations in the IBSFC.

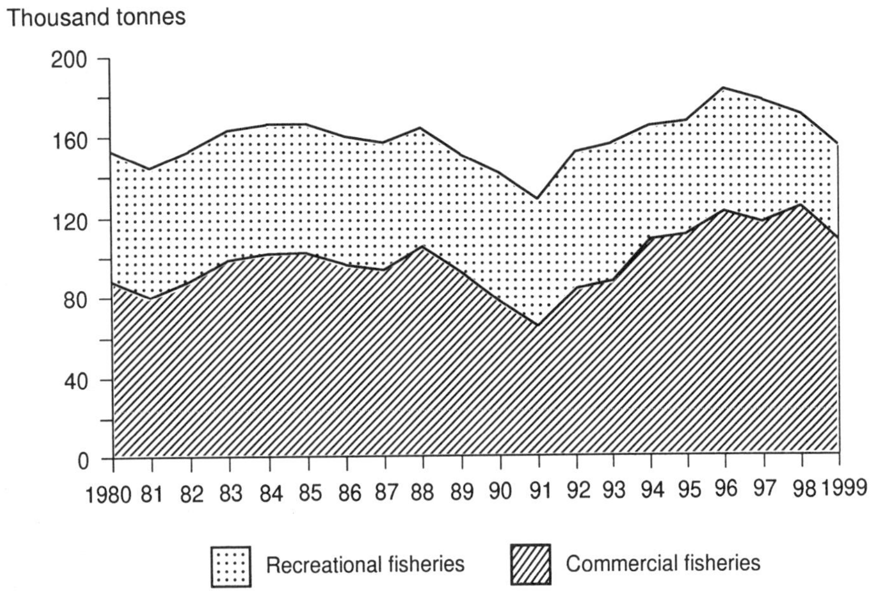

Figure 2.1 Commercial and Recreational Fisheries in Finland 1980-1999 (Source: Finnish Game and Fisheries Research Institute, 2000a)

2. PRIVATE AND PUBLIC WATERS

The definition of inshore waters is complicated by a distinctive pattern of ownership found in both Finland and to some extent in neighbouring Sweden. Until accession to the EU in 1995, Finnish territorial waters extended to only 4 nm from the shore. Although this limit was extended to 12 nm in line with the rest of the EU, the 4 nm limit retains some legal significance as regulations imposed by the IBSFC in respect of closed areas, for example, only have relevance outside the 4 nm zone.

Within the current 12 nm limits there are in effect two contrasting types of fishing waters in terms of ownership. Along with the lakes and rivers, the waters closest to the coast are in general privately owned: around the shores of the Gulf of Bothnia these waters are collectively administered by what originally were voluntary fishing associations based usually on the village, while in the south of the country around the Gulf of Finland the system of individual private ownership was less commonly organised according to village based associations. In the Archipelago Sea in south west Finland both systems of management occur. In 1902 the boundary between the privately owned coastal waters and the state owned, public waters was set at 500 metres beyond the 2 metre submarine contour with the effect that extensive areas of coastal waters were subject to private ownership and collective local management. So defined, privately owned waters will usually extend some 1-10 km from the shore or from the outermost islands. Following a reciprocal agreement Swedish fishermen also have access to fisheries within the 4 nm limits, subject to regulation and supervision by the Finnish authorities. This tradition of extending the idea of private ownership to include the neighbouring coastal waters is seen by Eklund (1994) to reflect the strength of influence of the peasantry on nation building in Finland.

Thus in Finland inshore fisheries can be defined in several different ways: by reference to the concept of territorial waters (12 nm); by reference to the pattern of ownership distinguishing between private and public coastal waters; or by patterns of fishing activity in which those who fish from their home port and return to land their catches within 24 hours are considered to be engaged in inshore fishing, irrespective of which category of waters they fish. In practice this last construct is probably the most relevant, though in terms of legal definition and management practice the ownership characteristics remain significant.

3. FINLAND'S INSHORE FISHERIES

3.1 Production

Two physical factors broadly determine the nature of fishing within the inshore waters. First, the low salinity of the Baltic Sea - ranging from 6-8 ppt in the southern areas of Finland to an average of 0-4 ppt in the Quark and the inner areas of the Gulf of Bothnia - defines the range of species found in Finnish waters. Second, the length and severity of the Baltic winter imposes a marked seasonal rhythm to fishing activity. In general, Finland's coastal geography can be characterised as an extensive but intermittent archipelago comprising countless islands and large numbers of small coastal communities. The largest archipelago area can be found lying between the Åland Islands and the south west coast of the Finnish mainland. It is here that the most extensive areas of privately owned coastal waters are to be found. Within the Archipelago Sea inshore fishing was formerly the principal means of livelihood; most of the inhabited islands and coastal communities could be described as fisheries dependent areas. In recent years, however the absolute values for fishing activity in terms of employment have fallen as a consequence of outmigration and the centralisation of economic activities typical of modernising regional economies.

The brackish nature of the coastal waters implies that fishing is directed towards a combination of marine (mainly herring, sprat, salmon and cod) and freshwater (perch, pike, whitefish etc) species; shellfish play no role in Finland's capture fisheries. But the dominant characteristic of fishing as an economic activity - its seasonality - is created by the sub-arctic climate. Finnish waters in general and inshore waters in particular are subject to varying degrees of ice cover each year. The highest rates of fishing activity naturally take place when the coastal waters are free from ice - broadly from April/May through to November. During this summer half of the year gillnets and various forms of traps are the most commonly used fishing gears in inshore waters, while trawling is the basis of the offshore sprat and herring fisheries.

Even during the winter season a certain level of fishing activity may be maintained under the ice cover using nets, hooks and traps. Figure 2.2 describes the seasonality of selected coastal fisheries, indicating the changing importance of different fisheries throughout the year and comparing these with the total catch for Finland's commercial fisheries in both inshore and offshore waters. The spring perch and pike fisheries are seen to overlap briefly with the summer and autumn whitefish fisheries, while the burbot is essentially a winter fishery. Table 2.1 indicates the use of

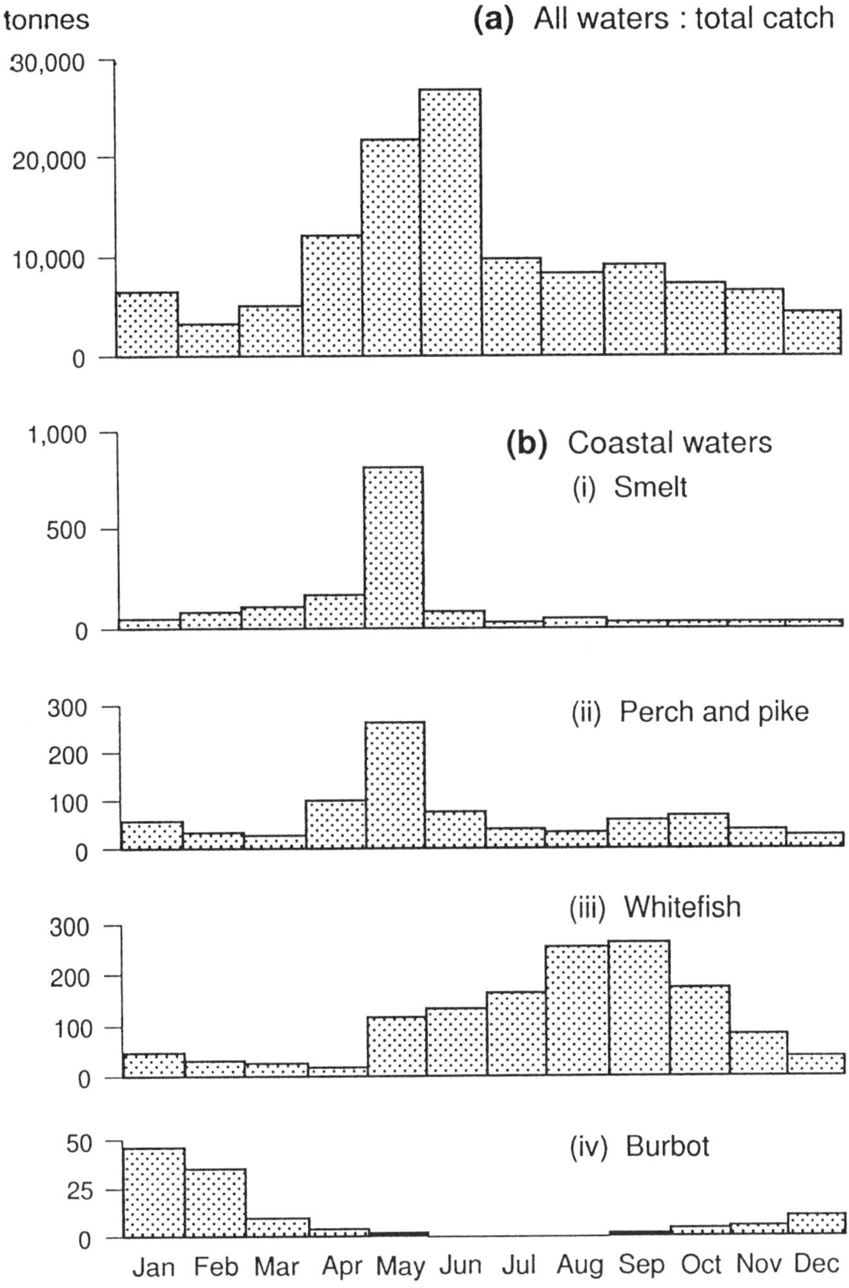

Figure 2.2 The Seasonality of Selected Coastal Fisheries in Finland (Source: Finnish Game and Fisheries Research Institute, 1997)

Table 2.1 Catches by professional fishermen in coastal fisheries, 1996

Species	Traps	Nets	Total catch	
	(number)	Catch days	Tonnes	FIM 000
Whitefish (powan)	506	120,623	1,280	15,922
Smelt	109	22,769	1,252	830
Salmon	1,031	1,535	975	12,527
Pike-perch	286	58,803	594	9,485
Perch	734	84,684	546	2,892
Pike	722	76,521	232	1,573
Trout	821	52,411	152	1,887
Burbot	530	29,576	111	1,101
Roach	328	14,737	107	111
Bream	400	23,188	102	292
Flounder	42	11,384	99	334
Vendace	157	1,489	86	712
Ide	200	16,315	22	26
Total	5,866	514,035	5,558	47,692

Source: Finnish Game and Fisheries Research Institute, 1997

Table 2.2 Catches by professional fishermen in all waters, 1999

Species	Landings		Value
	Tonnes	FIM 000	%
Offshore			
Herring	82,237	63,849	47.6
Sprat	18,886	6,652	5.0
Cod	1,569	13,417	10.0
Sub-total	102,692	83,918	62.6
Inshore			
Whitefish	1,246	18,583	13.9
Salmon	612	10,577	7.9
Pike-perch	438	7,752	5.8
Perch	821	4,412	3.3
Trout	103	1,598	1.2
Pike	278	1,936	1.4
Burbot	129	1,284	0.9
Sub-total	3,627	46,142	34.4
Others	1,385	4,052	3.0
Total	107,704	134,112	100.0

Source: Finnish Game and Fisheries Research Institute, 2000b

the two principal types of gear used by professional fishermen in the different coastal fisheries, while Table 2.2 demonstrates the relative importance of offshore and inshore fisheries to total production. Although the offshore species dominate the picture, accounting for 95% of the volume of landings and some two thirds of the revenue from the professional sector, the significance of the small scale inshore fisheries clearly lies in their targeting of high unit value species through what is in effect a labour intensive activity.

Aquaculture, principally for rainbow trout, has developed rapidly during the last 15 to 20 years providing the coastal communities and a number of fishermen with a valuable complement to their traditional fishing activities. It also meets the demands of the consumer markets for all year round supplies of fresh fish products. Production is concentrated in the coastal areas, especially in the archipelago of south west Finland and the Åland Islands.

3.2 The fishing fleet

Following entry to the EU in 1995 all commercial fishing boats in Finland had to be registered revealing for the first time the true extent and nature of the active - or potentially active - fishing fleet. In fact the number of boats (3,762) seems somewhat out of balance with the numbers of commercial fishermen (2,750); at least part of the explanation lies in the fact that many inshore fishermen own two or more boats under 8 m for use in different coastal fisheries. For the purpose of the EU's Multi-Annual Guidance Programme the fleet has been divided into different functional segments depending on vessel size and types of gears used (Table 2.3b). The picture which emerges again emphasises the overwhelming dominance of the small scale, inshore sector.

Table 2.3 The Finnish fishing fleet, 2000

a) Vessels by length (m)

Less than 6	6-7.99	8-11.99	12-20.99	21-27.99	28 and over	Total
1,687	816	1,031	179	39	10	3,762

Source: National Board of Navigation, Ministry of Agriculture and Forestry, 2000

b) Category

	Pelagic trawlers	Bottom trawlers	Passive gear	Coastal fisheries	Total
Number	207	3	76	3,476	3,762
Average GRT	47	150	26	3	55

Source: National Board of Navigation, Ministry of Agriculture and Forestry, 2000

3.3 Employment and the socio-economic structure of inshore fisheries

According to the register of active fishermen in Finland, at the end of 1997 employment in the catching sector was 2,750 with a further 5,200 persons employed in some 200 processing plants handling the produce of both marine and freshwater capture and culture fisheries. However, by far the greatest proportion of fishermen are classified as part time or non-professional (Table 2.4).

Although the figures do not distinguish between those employed in the inshore and offshore sectors, it is safe to assume that the vast majority of those earning less than 30% of their total income from fishing are in fact inshore fishermen; however, any attempt to maintain a clear separation between the two sectors will be thwarted by the significant numbers who combine a summer coastal fishery with participation in the offshore fisheries during the spring and autumn. Nonetheless, according to Varjopuro and Salmi (1999), it seems that rather more than half of all commercial fishing enterprises are solely dependent on the coastal fisheries, fishing for up to three or four months of the year and earning substantially lower incomes than their counterparts in the offshore fishery.

A further important inference to be drawn from Table 2.4 is that in Finland there are very few full time fishermen, in the sense that they rely wholly on fishing for both employment and income. This again is a clear reflection of the very strong imprint of the seasonality of Finnish fisheries on the socio-economic structure of the industry.

The distinction which is drawn between 'professional' and other fishermen is significant in a number of respects. In the first place, only those who derive more than 30% of their annual income from fishing can qualify for various forms of state subsidy, while those earning less than 15% from fishing are prevented from deploying certain types of commercial fishing gears outside privately owned waters. Tensions between 'professional' and 'other' fishermen exist over the allocation of access to resources and marketing behaviour. Casual and part time fishermen are often accused of

Table 2.4 Fishermen 1997 according to level of income (%) derived from fishing

	Professional fishermen		Non-professional	Total
	Over 30%	15-30%	Less than 30%	
Number	850	200	1,700	2,750
%	31	7	62	100

Source: Ministry of Agriculture and Forestry, 1997

undermining the quayside markets by their willingness to accept relatively low prices and their informal private sales of fish within the local community. In effect, the very legitimacy of the 'non-professional' fishermen's participation is being challenged.

To put this issue of legitimacy in a clearer perspective, it is necessary to distinguish between two contrasting types of part time fishing. First there are those mainly of urban origins in permanent waged (or salaried) employment for whom fishing is essentially a recreational activity; secondly, there are those for whom fishing is an important - indeed, potentially crucial - source of income and who usually combine fishing with jobs within the rural economy (farming, forestry, fur farming, construction work etc). Although the earnings generated through fishing in this second category may be low they are often integral to the household income strategy and may also be compensated by what is regarded as a highly valued way of life (Varjopuro and Salmi, 1999). The imposition of further restrictions on part time involvement could lead to an unravelling of the intricate patterns of pluriactivity that help to sustain some of the remoter coastal areas.

Modern management systems tend, not surprisingly, to favour professional fishermen and, in consequence, to limit the opportunities for others, thereby reducing the flexibility of necessarily pluriactive local economies. Salmi *et al.* (2000) have neatly encapsulated this issue by inferring that whereas part time fishermen may have only a nuisance value for fisheries management they often perform a vital role in maintaining the viability of the rural economy in coastal areas.

4. MANAGEMENT INSTITUTIONS

The four principal marine species - Baltic herring, sprat, cod and salmon - are managed on the basis of recommendations from the IBSFC. All commercial fishing activity concerning these species is therefore regulated in accordance with the provisions of the Common Fisheries Policy. By contrast, the exploitation of species which make up the inshore fisheries (see Table 2.1) are subject to regulations laid down by the national fisheries authorities and by the statutory fishing associations and regional organisations.

The basic legislation governing Finland's fisheries sector is contained in the Fisheries Act 1982 and its subsequent amendments. Accordingly, the Ministry of Agriculture and Forestry, together with the specialist Department of Fisheries and Game, has overall authority. However, the unique pattern of both private and public fisheries occurring within the traditional 4 nm inshore zone, outlined above, has meant the development of a complex institutional framework for the management of Finland's inshore fisheries involving both decentralisation and devolution.

Privately owned waters are subject to management by statutory fishing associations (SFAs). The Fisheries Act 1902 had facilitated the establishment of fishing associations on a voluntary basis. Half a century later, the Fisheries Act 1951 made these associations a statutory obligation. Today there are some 7,600 associations covering some 90% of Finland's inland lakes together with the narrow strip of coastal waters varying between one and ten km in width. Their membership ranges from two to two thousand shareholders (Sipponen, 1999) with each shareholder able to exercise a number of votes in proportion to the extent of his/her private holding. Varjopuro and Salmi (1999) describe the SFAs as a decision making system which may have been functional at the start of the twentieth century but one which had adapted relatively poorly to the changing conditions throughout the century. In particular they had failed to incorporate new fishing interests.

The Fisheries Act 1951 also made provisions for SFAs to merge to form larger, more effective management units, but the process of voluntary rationalisation proved too slow with the effect that the government was obliged to intervene in 1982 to create statutory fishing regions in order to encourage a more coherent system for the management of both natural and human resources (Sipponen, 1998). A total of 222 regions, incorporating the SFAs, were designated. The great majority were concerned exclusively with inland waters; only 45 (21%) of the regions actually included coastal waters. At the same time, the composition of the regional boards was broadened to include not only private rights holders and SFA members but also the representatives of commercial and recreational fishing associations.

The 1982 Act also laid the basis for the management of public waters within the 4 nm (and later the 12 nm) territorial zone with the establishment of eleven provincial Fishery Districts as part of the system for promoting regional development entrusted to the Employment and Economic Development Centres. Among their powers are the issuing of permits for static gears in the salmon fishery, the definition of minimum landing sizes for non-quota species, the determination of the opening and closing dates for fishing seasons and the establishment of protected areas.

Evaluation of the achievements of this regionalised system of management is somewhat lacking in enthusiasm. Although the institutional changes have seen a shift in decision making powers from individual owners and their protective associations to the formal organs of the state, Sipponen (1999) believes that they have been far less successful in their contribution to the management of coastal waters than in the case of the inland lakes where issues relating to the conservation of fish stocks have been less to the fore. The role of regional management in the coastal areas has been overshadowed by EU regulation of the principal marine species in terms of actual workload. The owners of private coastal waters have experienced difficulties in endorsing the regional approach, unlike their counterparts in inland waters: the very proliferation of SFAs, together with the size and complexity of their membership, has probably made effective cooperation at the regional level much more difficult in comparison with the inland areas where the number of actors tends to be much smaller.

Regional management of coastal waters should ideally balance the many different and sometimes conflicting interests occurring both within and between different territories (recreational and professional fishing opportunities; fishing opportunities in private and state owned waters; exploitation patterns in neighbouring regions affecting shared fish stocks, *inter alia*). By definition, the regional unit will initially take care of its own interests; subsequently, it may strive for a more integrated approach which would require cooperation throughout the whole coastal area. The latter is more in line with current management criteria in terms of sustainable development and the precautionary approach, but may well prove difficult to achieve in a regionalised framework.

It is only recently, with the decision in 1993 to remove the built-in guarantee of a majority of votes for the private owners in deliberations at the regional level, that a more balanced representation of interests has been created. As a result, the fishing regions constructed around the SFAs continue to play only a minor role in the management of coastal fisheries in Finland. Indeed, their lack of authority in the eyes of some of their members and the reluctance to use the powers available to them effectively has probably hampered the development of management in the coastal zone.

By contrast, the government's regional authorities - the Employment and Economic Development Centres - have attempted to exert a greater influence on the development of all sections of the inshore fisheries and their management. But their actions have also been affected by the difficulty in creating a genuine sense of partnership between the owners of the private waters, the user groups and the state authorities at the regional level.

5. MANAGEMENT ISSUES

The broad thrust of policy within the fisheries sector overall is to maintain and develop the opportunities for fishing while adjusting the balance of fishing capacity to the level of sustainable resources allowing for the modernisation of the fishing fleet, and increasing the value of the final product through improved quality on landing and in processing. No distinction is made in the fulfilment of these objectives between inland, inshore and offshore fishing activities. From the consumers' perspective the future lies in the guarantee of high quality products of certifiable standard created under sustainable conditions. Long term planning, a precautionary approach and sustainable fisheries are key concepts in implementing management strategies for the Baltic under the banner of 'Baltic 21': they are expected to guide management decisions at international, national, regional and local levels (Baltic 21, 1998).

Confronting these broad goals are a number of existing or emerging issues which, if left unresolved, could constrain development and distort the intended trajectory. None of these issues are unique to Finland but some attain a particular significance in the special circumstances surrounding the nature of Finland's rural economy and the prevailing ecological conditions of the Baltic. They may also have a significant bearing on the future of the inshore sector in particular. A distinction can be drawn between those issues which are internal to fisheries management and relate to possible inconsistencies within the broad strategies for achieving sustainable development, on the one hand, and those which are concerned with external relations between the fishing industry and other forms of resource use in inshore waters (Table 2.5).

Table 2.5 A SWOT analysis of inshore fisheries in Finland

Strengths • exploitation of sustainable resources • high value species • caught and sold fresh • low cost systems of production • labour intensive systems • proximity to home ports • complement to flexible pluriactive local economies	Weaknesses • seasonality of production • small unit volumes of production • highly dispersed landing points • weak infrastructure • remoteness from urban markets • increasing numbers of unregistered fishermen • complex patterns of administration and management
Threats • potential impacts on ecosystem • overprotection of target/non-target species • external impacts on marine environment and ecosystems (pollution from industry, agriculture, forestry and aquaculture) • impacts from aquaculture (markets) • greater economic attraction from recreational interests • weak political influence of coastal community	Opportunities • integration of small scale fisheries in development of sustainable marine ecosystems • better utilisation of resources (quality) • expanding markets • low risk: low investment enterprises • more effective systems of co-management

Tensions exist within the inshore sector largely as a result of the unbounded increase in pressures from coastal tourism and the evolving status of recreational fishing. Underlying this situation are two related but strongly divergent trends. The first is the decline in the level of interest in small boat fishing especially among young people; this is linked, in part, to the increasingly unfashionable images of a pluriactive rural economy which run counter to modern concepts of the division of labour and job specialisation. On the other hand, the increasing centralisation of employment opportunities and community services is associated with rapid urbanisation of the population which, among other things, has generated a growing interest in recreational fishing. Thus, while there are fewer than 3,000 professional fishermen in Finland today, as many as 600,000 Finns have taken out state fishing licences and recreational fishermen's associations have a combined membership of between one and two and a half million.

Recreational fishermen are especially active during two particular periods of the year: during the summer vacation months when many urban dwellers take up seasonal residence in their second homes along the shores of the coast and lakes, and again in the spring (March-April) associated with ice fishing (*pilkfiske*) for certain freshwater species. Recreational opportunities are more or less evenly distributed throughout Finland but, with more than half the population concentrated in the large cities of the south of the country, conflicts between recreational and professional fishermen and between recreational fishermen and the owners of private fishing waters are most frequent along the southern coasts. Significantly, the growing political power of the recreational fishing lobby has already forced the Finnish parliament to take account of their demands for better access to privately owned waters in new legislation relating to angling and the use of hand gears.

A second cause for concern has been the recent growth of aquaculture to provide fish for restocking and enhancement schemes (e.g. salmon, seatrout, whitefish, pike-perch), to produce freshwater crayfish for rearing or consumption, and in the farming of finfish (mostly rainbow trout). Only the last of these activities directly affects the coastal areas. Mainly on account of emerging environmental problems in the mid 1980s, the aquaculture sector has been forced to look for new locations. Today the largest concentrations of coastal aquaculture enterprises occur in the archipelago areas of the Åland Islands and the south west. When situated in shallow waters, eutrophication problems can be very pronounced, especially at the local level, and where the aquaculture enterprises coincide with important spawning and feeding grounds for commercial wildfish species the impacts can seriously diminish the commercial fishing potential. Aquaculture also encounters problems with fish diseases in the normal course of events and there are risks of transmission of disease from caged to wildfish populations. Although pollution problems from aquaculture are dwarfed by those arising from agriculture, forestry and land drainage, the environmental impacts of aquaculture have become a general focus of concern in the Baltic (Eklund, 1996). The impacts are, however, greatest at the local level where there are obvious conflicts of interest with other activities involving both permanent residents in the coastal communities, seasonal residents and tourists.

Tourism is, in fact, a particular feature of the archipelago area and the generation of boating activities and shipping movements associated directly or indirectly with tourism is becoming a major problem for the management of the coastal zone, including issues of water quality, navigation and the designation of protected areas for fishing or marine wildlife conservation. The ferry traffic between Sweden and Finland - much of it passing through the archipelago areas - is huge, while the size and speed of modern ferries can create wave effects with the potential to disturb fishing activities using fixed gears and cause damage to the ecosystem. Moreover, the shallow nature of parts of the Bothnian Sea - together with the continuing effects of isostatic readjustment since the last ice age - means that navigation channels for both large and small boats must be continually maintained through dredging with inevitable impacts on marine habitats and local ecosystems.

In some ways, Finland can count itself fortunate not to have any significant proven reserves of oil and gas beneath her seas. On the other hand, the need to develop and maintain road communications especially in the archipelago area during the 1960s and 1970s caused many hundreds of kilometres of embankments to be constructed, impacting locally on patterns of water flow, water quality and fish stocks. Likewise the exploitation of sea sand in certain parts of southern Finland has had negative effects on coastal spawning and feeding grounds, creating oxygen free, sterile areas of seabed. Further problems also arise as the result of using coastal waters for the discharge of waste waters from a variety of urban, industrial and agricultural sources. Where poorly functioning urban sewage systems discharge directly into the sea, the pollution load may be considerable, though this is now a relatively rare occurrence. A potentially much greater threat comes from the discharge of rivers which have accumulated a pollution load from both agricultural and forest areas in their catchments.

Looking to the future it is possible that the most serious constraints on the economic and social sustainability of commercial inshore fisheries in Finland could come from the growing concerns over the state of the marine environment in the Baltic in general and the conservation of marine wildlife in particular. Conservation management is largely divorced from fisheries management both at national and European levels. Development of the

Natura 2000 network of protected sites covering both terrestrial and coastal environments, combined with existing legislation to protect environmentally valuable tracts of coastline, could make it increasingly difficult to secure essential infrastructural developments (landing points, harbours, storage facilities and processing plants etc). Fishing may be directly affected by proposals to protect certain species such as the Baltic seal population and perhaps more significantly by concern over the state of the wild salmon stocks in the Baltic.

Action needs to be taken now to ensure that the public in general, and the conservation lobby in particular, appreciates the difference between inshore and offshore fishing activities in terms of their impacts on the marine ecosystem. At the same time every effort should be made to guarantee that inshore fisheries in Finland adhere to the principles of responsible fishing and sustainable fishing practice. Only through such actions can the climate of misunderstanding between conservationists and fishermen be replaced by a sense of common purpose in achieving a sustainable future.

6. REFERENCES

Baltic 21 (1998) *An Agenda 21 for the Baltic Sea Region*, Baltic 21 Series, No. 1. ISSN 1029-7790.

Eklund, E. (1994) *Kustfiskare och Kustfiske i Finland under den Industriella Epoken, Studies i en Yrkesgrupps yttre Villkor, Sociala Skiktning och Organisation 1860-1970*, SSKH (Svenska Social-och Kommunal Högskolan) Skrifter 5, Helsingfors.

Eklund, E. (1996) Aquaculture in the Baltic Sea: regional development and environmental conflict, in C. Bailey, S. Jentoft, and P. Sinclair (eds.), *Aquacultural Development: Social Dimensions of an Emerging Industry*, Westview Press, Boulder Colorado, pp. 59 -67.

Finnish Game and Fisheries Research Institute (1997) *Professional marine fisheries, 1996*, SVT Official Statistics of Finland: Agriculture, Forestry and Fishing 1997: 8, Finnish Game and Fisheries Research Institute, Helsinki.

Finnish Game and Fisheries Research Institute (2000a) *Finnish fisheries statistics - kalatalous tilastoina*, Finnish Game and Fisheries Research Institute, Helsinki.

Finnish Game and Fisheries Research Institute (2000b) *Professional marine fisheries, 1999*, SVT Official Statistics of Finland: Agriculture, Forestry and Fishing 2000: 7, Finnish Game and Fisheries Research Institute, Helsinki.

Ministry of Agriculture and Forestry (1997) *National register of professional fishermen*, Ministry of Agriculture and Forestry, Helsinki.

National Board of Navigation, Ministry of Agriculture and Forestry (2000) *National fleet register for professional fishermen*, National Board of Navigation, Ministry of Agriculture and Forestry, Helsinki.

Salmi, P., Salmi, J. and Lappalainen, A. (2000) Finnish part-time fishery: a flexible strategy or a nuisance for fisheries management, in D. Symes (ed.), *Fisheries Dependent Regions*, Blackwell Science, Oxford, pp. 132-142.

Sipponen, M. (1998) Fisheries regions - a tool for cooperation in fisheries management? in P. Hickley and H. Tompkins (eds.), *Recreational Fisheries: Social, Economic and Management Aspects*, Blackwell Science, Oxford, pp. 263-278.

Sipponen, M. (1999) Fisheries regions: an organisational structure for fisheries management in Finland, in D. Symes (ed.), *Alternative Management Systems for Fisheries*, Blackwell Science, Oxford, pp. 51-59.

Varjopuro, R. and Salmi, P. (1999) The functionality of fisheries management from the perspective of commercial fishermen. Two cases in comparison, in D. Symes (ed.), *Alternative Management Systems for Fisheries*, Blackwell Science, Oxford, pp. 93-105.

Chapter 3

SWEDEN

Gunnar Thoresson

1. INTRODUCTION

Sweden has only been part of the European Union since 1995 and the administration is still engaged in a process of adaptation to meet the particular demands and opportunities created by membership. Several analyses (see, for example, Anon, 1998) have attempted to assess the efficacy of Sweden's current fisheries system from a European perspective. No reasons have been identified for not continuing to operate within the framework of the existing management model, though a recent government bill (1997) has indicated that certain specific areas may need to be further developed in order to comply more fully with the requirements of the Common Fisheries Policy (CFP).

Rules and regulations concerning fisheries management are applied according to specific geographical boundaries. Thus the CFP applies to marine waters outside the baseline and only in respect of commercial, not recreational, fisheries. National legislation governs the conduct of fisheries within the territorial waters, that is within the 12 nm limits. The fisheries department of the Ministry of Agriculture, Food and Fisheries, together with the National Board of Fisheries, is responsible for the application of fisheries law and the implementation of EU regulations. In effect, the National Board has exclusive authority concerning the performance and management of the inshore fisheries. Provincial departments may also undertake certain responsibilities as delegated by the National Board and share responsibility for others with the National Board.

According to Swedish law, waters out to 4 nm from the baseline are defined as 'coastal'. Within this area there is an exclusion zone for trawl fisheries up to 4 nm from the baseline in the Baltic Sea, 2 nm in the

D. Symes and J. Phillipson (eds.), Inshore Fisheries Management, 43–59.

Skagerrak and 1 nm in the Kattegat, subject to local exemptions. Coastal fisheries are therefore defined legally according to both geographical limits and types of fishing activity, depending on the objectives. Meanwhile, the Institute of Coastal Research - a part of the National Board of Fisheries - classifies all fisheries within the 4 nm limits, with the sole exception of trawling by vessels over 12 m in length, as coastal fisheries irrespective of the types of gear. It is this last definition which is used in the following account of Sweden's inshore fisheries.

2. SWEDEN'S INSHORE FISHERIES

2.1 Environmental background

Including the islands and fjords, Sweden has a coastline of more than 25,000 km extending from the innermost parts of the Gulf of Bothnia in the north to the shores of the Skagerrak in the west. Its fisheries are regionally diverse, with the species composition reflecting major differences in the key elements of the marine environment, namely salinity and temperature. Salinity varies from less than 3 ppt in the northernmost reaches of the Bothnian Bay to 'normal' marine conditions of around 25 ppt on the Skagerrak coast. Low salinity prevents the reproduction of all commercial shellfish species in the Baltic. Among the flatfish species only flounder and turbot are able to reproduce successfully in northern parts of the Baltic. However, most of the region's commercial freshwater species will spawn in the brackish coastal waters of the Baltic.

Variations also occur in recruitment capacities and growth within the coastal waters due to different topographical conditions, the availability of suitable substrates, the presence of sheltered areas and different temperature regimes. All of these variations are reflected in the distinctive regional patterns of fishing activity (see below). Commercially, the most important marine species found in coastal areas are cod, herring, *nephrops*, eel, lobster and, to a growing extent, turbot. The commercial freshwater species include salmon, sea trout, vendace, whitefish, perch, pike and pike-perch.

2.2 Employment, production and the fishing fleet

In the 50 years between 1945 and 1995 the number of professional fishermen fell from around 16,000 to 3,300 - a decline of almost 80% - and to judge from the current age structure of those employed in the fishing industry, this trend seems destined to continue. The areas where the reduction has been most pronounced include the northern and eastern coasts fronting on to the Baltic and the more rural parts of the west coast away from the influence of Göteborg, Sweden's major deep sea fishing port. Much of this pattern of decline can be attributed to the processes of modernisation occurring within the fishing industry and to the contrasting perceptions of young people with regard to employment in the inshore and offshore sectors (Píriz, 2000). Thus, while recruitment to the capital intensive offshore sector based in and around Göteborg has been fairly buoyant of late, young people have increasingly turned their backs on the prospect of employment in the small scale, rurally based inshore fisheries. As a result the active fishing population has aged quite markedly: in 1985 the average age for fishermen throughout Sweden was 40; only 13 years later it had risen to 48 (Píriz, 2000). Locally the average age may be very much higher. Not surprisingly, the fishing fleet has contracted considerably during the last 25 years (Table 3.1). The rate of decline has tended to quicken and has been most marked in the smallest size classes. In the space of five years before entry to the EU, the registered fleet had fallen from 4,358 vessels to 2,742 (-37%), and for boats under 8 m numbers more than halved over the same period.

Despite the obvious contraction in the inshore sector, it remains numerically dominant in terms of both employment and vessel numbers (Table 3.2). In the mid-1990s, vessels under 12 m accounted for almost three quarters of the registered fleet and for *circa* 60% of Swedish fishermen. However, its contribution to overall production was slight. Around 10% of total landings by volume came from the under 12 m sector; the most important species were herring, sprat, cod vendace, eel, whitefish, salmon, *nephrops*, pike, perch and flounder - a mix of marine and freshwater species.

Table 3.1 Sweden's fishing fleet, 1970-1995

	Year					% change
	1970	1979	1985	1990	1995	1979-95
Less than 12 m	-	6,465	4,708	3,828	2,272	-65
More than 12 m	-	532	527	530	470	-12
Total	7,969	6,997	5,235	4,358	2,742	-61

Source: based on Hultkrantz *et al.*, 1997

Table 3.2 Sweden's fishing fleet, 1997 by vessel size (GRT)

	Less than 25	25-49.9	50-99.9	100-149.9	150-249.9	250-499.9	500 and over	Total
Number	1,933	95	85	34	39	35	5	2226
Total GRT	7,532	3,379	6,163	4,284	7,726	11,645	3,234	43,963

Source: based on Gustavsson, 1998

Herring and sprat contributed 9,500 tonnes, almost two thirds of the total landings by volume; the greater part (75%) was intended for human consumption. In terms of value the more important species were *nephrops* (1,100 tonnes valued at 60 million SEK) and eel (900 tonnes valued at 50 million SEK).

2.3 Recreational fisheries

According to a national survey in 1999-2000 around 2.3 million people - or 37% of the adult Swedish population aged 16-74 - had been fishing on at least one occasion during the previous 12 months. The catch from the recreational fishery was estimated at close to 58,200 tonnes (Bengtsson, 2000); of this 34,500 tonnes (60%) was taken in inshore waters. Herring, freshwater species (pike, perch, pike-perch and burbot) together with eel accounted for *circa* 8,000 tonnes, gadoids a further 8,000 tonnes, followed by flatfish (4,000 tonnes) and salmonids (4,000 tonnes). Of all those participating in the so-called recreational fisheries, roughly three quarters could be described as 'sports fishermen' using mainly rod and line; the remaining one third were 'household fishermen' deploying gillnets and other fixed gears to catch fish principally for home consumption though it is likely that small private sales of fish would also result. Local surveys conducted by the Institute of Coastal Research at various localities around the Swedish coast suggest that the national sample survey may overestimate the role of the recreational fisheries.

2.4 Aquaculture

Swedish aquacultural production serves a number of different purposes: the farming of finfish and blue mussels for human consumption; the rearing of fish to compensate for losses in recruitment caused by the disruption of river regimes through the construction of dams for hydroelectric power generation; the rearing of fish to support sports fishing activities. Around 60% of the finfish cultivation, together with all mussel production, take place in coastal marine waters with the remainder in freshwater locations.

Total finfish production for consumption in 1997 amounted to 5,200 tonnes valued at 127 million SEK (Gustavsson, 1998). Of this, rainbow trout (4,800 tonnes) was by far the most important species; others included eel, salmon, trout and arctic char. Mussel production in 1997 was in the region of 1,400 tonnes. Compensatory rearing of salmon and trout involved the release of over three million fry, mainly in rivers flowing into the Baltic (Olsson, 1997).

3. REGIONAL PATTERNS

Sweden's coastal fisheries can be divided into three main regions: the Gulf of Bothnia comprising ICES Division IIId (sub-areas 30 and 31); the Baltic proper (ICES Division IIId (sub-areas 24, 25, 27 and 29N); and the west coast, including the Sound separating Sweden and Denmark (ICES Divisions IIIa, b). These regions are indicated in Figure 3.1. The information on catches and landings given below is derived from official statistics which do not necessarily reflect the totality of commercial fisheries in Sweden. There are, for example, good reasons to believe that inshore catches are often considerably underestimated, especially as the unlicensed fishing which takes place in private waters is not fully recorded.

3.1 The Gulf of Bothnia

Only 15% of Sweden's small boat fleet (less than 12 m) is located in this region and the number of professional fishermen has been more or less constant since the beginning of the 1980s. Only a few species are exploited commercially - mainly vendace, whitefish, salmon, trout and herring. In the northern part of the Gulf - in the area known as the Bothnian Bay - vendace is the most important species in terms of volume and value. The status of vendace stocks is uncertain and the National Board of Fisheries currently recommends a reduction in fishing effort. Vendace is targeted mainly by an autumn trawl fishery directed towards the roe, which accounts for between 75% and 80% of the total catch value, with the bulk of the catch intended for reduction. Catches, amounting to only 100 tonnes during the first half of the 1960s, exceeded 1,000 tonnes per annum at the start of the 1990s but have fallen back to 400-600 tonnes per annum in more recent years. The vendace trawl fishery is regulated through vessel permits granted only to fishermen in the two most northerly counties of Norrbotten and Västerbotten for a three year period. During the 1990s the number of licences was cut from 60 to 48. In the year 2000 management of the vendace fishery was undertaken by the fishermen themselves in a programme which is to be evaluated during the period 2001-2003. Recreational fishing accounts for around a fifth of the annual catch.

Salmon is the second most important species in the region, where it is caught during its return migration to the spawning rivers mainly in large trap

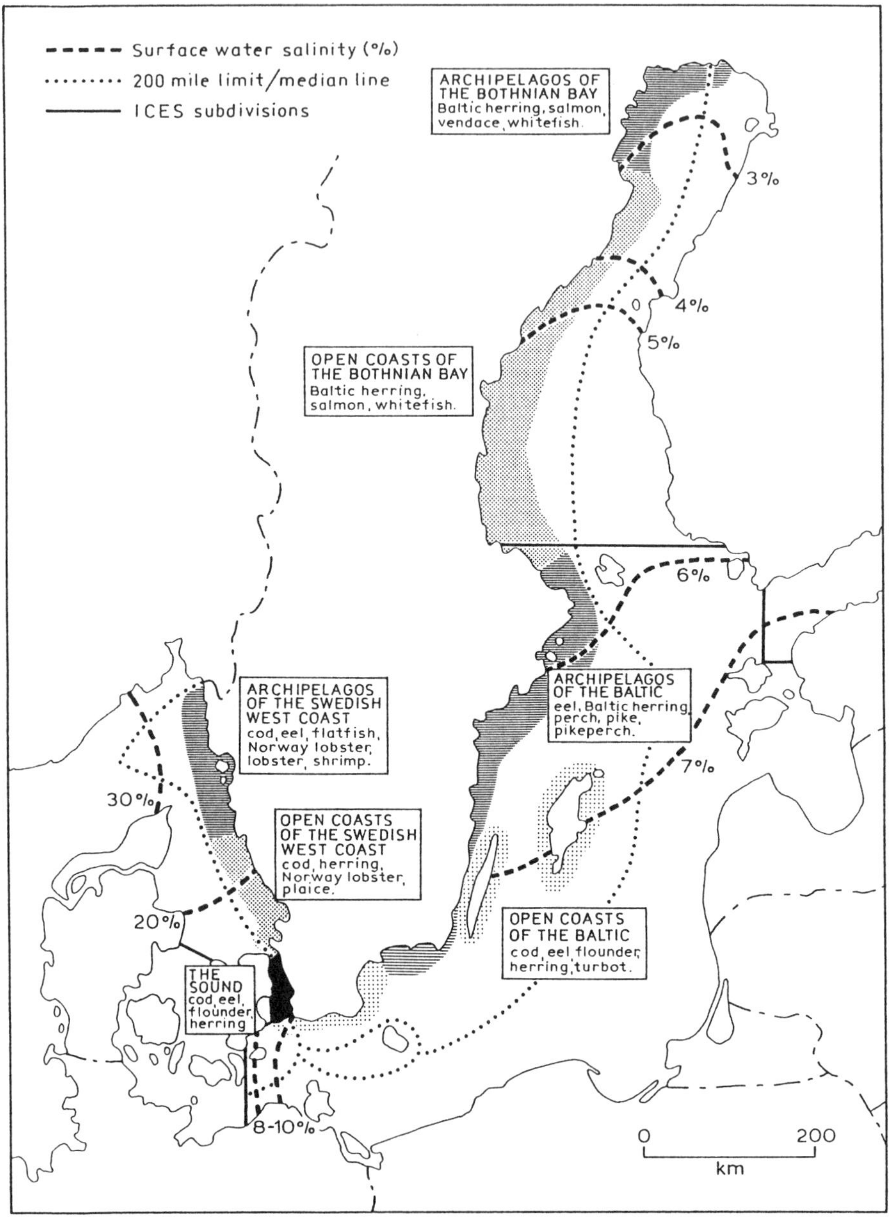

Figure 3.1 Sweden's Inshore Fisheries

nets. In 1997, 30% of the national total allowable catch (TAC) (112,600 fish) was caught in the Bothnian Bay area. At present the fishery has to rely for up to 90% of the catch on the restocking of the rivers with hatchery reared fry intended to compensate for losses to natural stocks caused by hydroelectricity development. A new management programme intended to conserve and restore wild salmon stocks has been proposed by the National Board of Fisheries. The programme is based on the International Baltic Sea Fisheries Commission's (IBSFC) *Salmon Action Plan, 1997-2010*; by the year 2010 it is intended that half of the salmon production capacity should come from rivers with wild salmon stocks. Restoration of the stocks and rights of access to wild salmon stocks remain controversial issues within the Swedish industry and form recurrent topics for debate in the Swedish parliament.

A third important species for the inshore fisheries is whitefish. Commercial landings have decreased from around 600 tonnes annually at the beginning of the 1950s to around half that level today. Whitefish is caught along the whole of the east coast with gillnets and large trap nets. There are two forms - a small sea spawning and a faster growing river spawning variety, with the latter caught mainly during the autumn spawning migration. Landings in 1996 were 170 tonnes (valued at 2.1 million SEK). While several stocks of the migratory, river spawning whitefish are on the brink of overexploitation, fishing effort in respect of the sea spawning variety could be increased.

Herring was once the backbone of the Gulf of Bothnia's coastal fisheries and a major national food resource at least until the 1940s. Since 1945, however, the market has consistently declined. Today industrial fishing for herring is principally carried out with trawl in the open sea, while the inshore catches using gillnets and large trap nets are mainly used in the production of cured herring products. The total Swedish catch in the Gulf of Bothnia in 1995 was only 500 tonnes.

Recreational fishing is relatively important in the region especially in relation to freshwater species like perch and pike and also whitefish. A regional survey conducted in 1991 attributed shares of between 70% and 95% of the total catch to recreational fishing in the case of freshwater species like perch, pike, pike-perch, burbot and trout; and in the order of 60% for whitefish (Sandström, 1994).

3.2 The Baltic Proper

The principal species for the inshore fisheries is eel - mainly the migrating silver eel taken with pound nets - but also the resident yellow eel mainly caught with fyke nets. The eel catch has fallen by 60-70% since the 1950s to a level of around 600 tonnes in the early 1980s where it has since remained.

On a seasonal basis the eel fishery is combined with gillnetting for freshwater species (perch, pike, pike-perch and whitefish) or marine species (cod, flounder and, to a lesser extent, herring). It is largely limited to privately owned waters where the fish is taken either by the owners themselves or by other fishermen who rent the fishing rights from the owners. Eel fishing may be combined with other non-fishing related employment; part time and recreational fishing with commercial gears are common in private waters. Studies by the Institute of Coastal Research indicate that private ownership of inshore waters in the Baltic has tended to restrict the development of economically viable enterprises by restricting access to professional fishermen.

A second important coastal fishery is based on the Baltic cod caught mainly with gillnets. Cod fishing is commonly combined with seasonal participation in the flounder, turbot and herring fisheries. The cod fishery is subject to international regulation through the IBSFC and the EU and the resource is shared somewhat unevenly between the small scale inshore fishery and the large scale capital intensive sector. In 1996 some 400 boats under 12 m landed 8,000 tonnes, thus accounting for roughly 25% of the Swedish TAC for Baltic cod. Cod stocks in the Baltic are presently under considerable pressure due to the overcapacity of the fleet; as a consequence the economic returns from the fishery have been in decline.

A small fraction of the Baltic inshore fleet, located mainly in a handful of fishing communities on the larger islands and in the southern parts of the region, has focused its activity on other target species such as flounder, turbot, salmon and herring.

Altogether the region's fisheries involve some 1,200 professional fishermen, the majority of whom are engaged on a full time basis. Of these between 850 and 950 can be classed as operating within the coastal fishery. Some 300 or so rely mainly on the eel fishery and the remainder on cod and other species. In 1996 the licensed fleet of 760 vessels was dominated by inshore vessels under 12 m. However, to this figure there must be added an unknown but significant number of small boats used by either recreational fishermen or by full time but unlicensed operators working exclusively in the privately owned coastal waters. Again, as in the Gulf of Bothnia, a substantial part - and for some species like perch and pike the dominant part - of the total catch is taken by the recreational fishermen.

3.3 The West Coast

By comparison with the fisheries in the Gulf of Bothnia and the Baltic, the west coast is rather more strongly oriented towards the capital intensive offshore sector, with Göteborg as the principal deep sea port. This is the area with the highest levels of basic production and the greatest biological diversity, giving rise to a wider variety of coastal fishing patterns than elsewhere in Sweden's coastal waters. Coastal fisheries on the west coast can be divided into two broad groups: finfish and shellfish. Herring and cod dominate the catch in terms of volume, with the few purse seiners accounting for roughly half the total inshore catch. Both species have been subject to declining stock and catch levels in recent years. But the most important species by value is *nephrops*. Although there had been an expanding pot fishery in the 1980s, around 80% of the catch is taken by bottom trawls operated from relatively small boats fishing - under exemption - within the trawl-free 2 nm zone. Other important shellfish include lobster and crab. Eel remains an important target species along the length of the west coast: a substantial part of the catch consists of undersized eel used for restocking fisheries on the east coast and in inland waters.

As elsewhere in Sweden, there has been a substantial reduction in the numbers of both fishing vessels and active fishermen in recent decades. Between 1970 and 1995, the number of boats under 8 m fell by up to two thirds to a level of 700 boats while over the same period the numbers of professional fishermen has more or less halved.

4. FISHERIES MANAGEMENT

4.1 Fishing rights

Together with Finland, Sweden shares the distinction - unique in Europe - of having part of its coastal waters in private ownership. As defined in the 1950 Fisheries Act, such private waters extend to within 300 metres of the shoreline of the mainland or of islands over 100 metres in length or alternatively occur inside the 3 metre submarine contour if this is beyond 300 meters from the shore - an area appreciably smaller than that in Finland (see Chapter 2). Otherwise waters in the Swedish archipelago, connected to the open sea by sounds less than 1 km in width, are classed as private waters. Despite the narrowness of the zone, the concept of private waters has considerable economic and social ramifications for the inshore fisheries. In Sweden, the principle of *allemansrätt* does not apply to fishing rights; although the public may exercise certain rights of access to privately owned land and water, the rights to fish are reserved exclusively to the owner. Thus the individual owner is solely responsible for managing the access to, and use of, fish resources within his/her property - though serious abuse of the rights and responsibilities can render the owner liable to legal action by the appropriate public authority. The exclusive fishing right can be exercised whether or not the owner holds a fishing licence. Moreover, the fishing rights can be transferred temporarily to other persons on purchase of a fishing pass, though this practice does not apply equally throughout the length of Sweden's coastline nor in respect of all species. On the north east coast, for example, salmon are excluded from the transfer rules, while on the west coast - including the whole of Skåne - private fishing rights were bought out by the state over a century ago. Therefore, with the exception of the oyster beds, all waters in the region are now regarded as public property.

Beyond the outer limits of the private waters, fisheries in public waters are administered by the Ministry of Agriculture, Food and Fisheries in conjunction with the National Board of Fisheries. Since 1968, however, in line with international law, Sweden's territorial limits have been extended from the traditional 4 nm limits to 12 nm. Five years later, in 1973, the median line principle was adopted in the definition of Sweden's national waters.

4.2 Organisational structures for inshore management

Since the 1940s - and largely coincident with the setting up of the National Board of Fisheries (*Fiskeristyrelsen* and, from 1991, *Fiskeriverket*) - the role of the state in the management of both professional and recreational fisheries has been progressively strengthened. Today the management system involves a complex, hierarchical structure of national, regional and local bureaucratic organisations in which the National Board, the County administrations, the coastguard service and, to a lesser extent, the local municipalities are the principal players (see Fig. 3.2). Also involved with specific responsibilities are the Environment Protection Agency (*Naturvårdsverket*), the Juridical Board for Public Lands (*Kammarkollegiet*), the Board of Agriculture (*Jordbruksverket*), the National Food Administration (*Livsmedelsverket*) and the National Veterinary Institute (*Sveriges veterinärmedicinska anstalt*).

The National Board of Fisheries is the lead agency for fisheries management with statutory responsibility for the sustainable use of aquatic resources. It collects landing statistics, negotiates the allocation of national quotas, defines the opening and closing dates for fishing seasons, issues commercial fishing licences and distributes grants and loans to the industry. Contact with the industry is achieved largely through the County administrations which assume certain delegated responsibilities for the implementation of national policy. The functions of the County administration are to safeguard the interests of fishing and conservation of

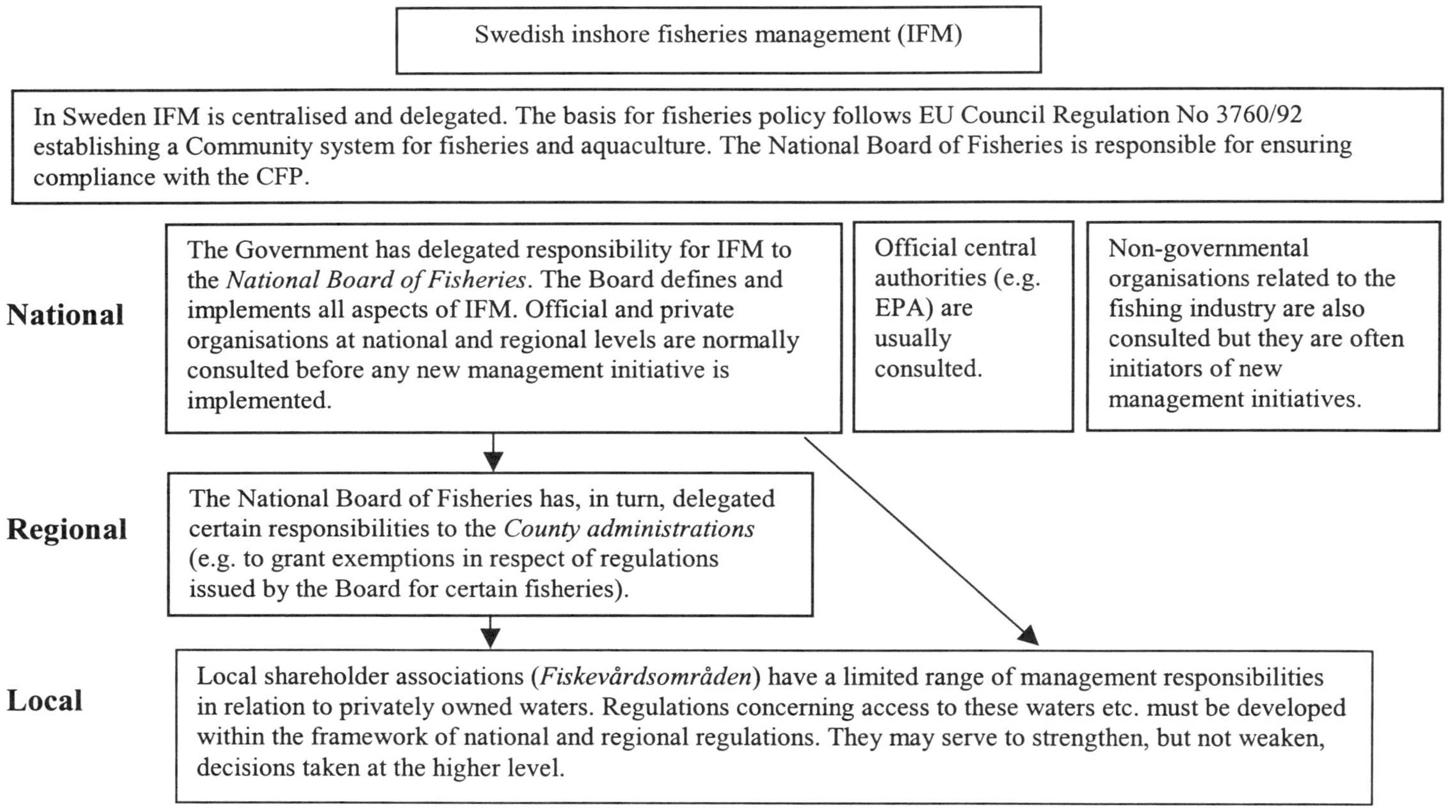

Figure 3.2 The Institutional Framework for Fisheries Policy in Sweden

the stocks, provide information to the industry on current regulations, handle applications for loans and subsidies from the EU, allocate professional fishing licences and permits for sports fishing and fish farming, as well as undertaking analyses of fish populations and water quality.

Proposals for changes to the systems of regulation may be initiated by local fishermen, the owners of private waters or by the various levels of administration; such proposals are subject to scientific, that is biological, assessment, though rarely with the additional support of economic or social impact studies. Ultimately, however, the National Board remains the final arbiter for all decisions relating to fisheries management within the 12 nm zone, subject to the approval of the Ministry.

The fishermen's associations also play an important role in defining the allocation of national quotas. Individual stakeholders are represented through professional associations such as the Swedish Fishermen's Federation (*Sveriges Fiskares Riksförbund* or SFR), the Association of Fishing Water Owners (*Sveriges fiskevattenägarförbund*) and the National Anglers Association, together with a number of regional organisations, of which the West Coast Fishermen's Association (*Svenska Västkustfiskarnas Centralförbund* or SVC) - strongly identified with the large scale herring fleet - is the most influential.

In effect, the most important linkage within the management system is that which has developed between the National Board and the SFR. Nominally a non-political association, SFR was founded in 1948 and today represents some 2,700 members or roughly 80% of the registered fishermen through its five regional branches. According to Byron (1993), the SFR is little more than a quasi-governmental organisation created in an attempt to impose a centralised hierarchical form on the means of representation of fishermen's interests in Sweden. Together with the National Board, it has succeeded in focusing attention on the maximisation of economic opportunities, improvements in the quality of production for both domestic and export markets and on the professionalisation of the fisheries sector. As in other European countries, however, the small boat inshore fishermen have sensed a marginalisation of their particular interests in the management process. Attempts to establish professional organisations dedicated to advancing the interests of inshore fishermen have met with little success.

Formerly, the first hand sales of fish from both the inshore and offshore sectors were organised by the fishermen's own local associations, with the benefit of state support. This role is now being superseded by the three producers' organisations (POs) created as a consequence of accession to the EU and responsible for managing the sales of their members' catches within a stabilised price regime. Unlike POs in some other Member States, in Sweden their activities do not extend into the realms of quota management.

Together the POs have a combined membership of about 1,300 and handle around 70-80% of commercial sales. The largest of the three (*Sveriges Fiskares* PO) is essentially a national organisation with a membership of 1,100 boats, ranging from the largest to the smallest categories; the second and third POs handle the small boats in the Gulf of Bothnia and the west coast trawl fishery respectively. There is a separate producers' organisation for the aquaculture sector.

5. REGULATORY SYSTEMS FOR INSHORE FISHERIES

The system of inshore fisheries management targets both selected species and specific types of fishermen. The distribution of many commercially important species overlaps inshore and offshore waters and these are therefore subject to regulation through the IBSFC and the EU's Common Fisheries Policy. For herring and cod, two of the most important 'transboundary' stocks, quotas were introduced in 1981 and 1989 respectively; salmon quotas were brought into force in 1991. Today, recommendations for the regulation of cod and herring are made through the IBSFC - to which the EU is a party - and implemented through the terms of the CFP. In the case of salmon, the implementing agency is the coastal state. Most other species of relevance to the inshore fisheries are largely unregulated: the exceptions are vendace in the Gulf of Bothnia and lobsters on the west coast.

Of somewhat more direct concern is the regulation of inshore fishermen according to their professional status. In Sweden, official designations recognise two broad categories: professional and recreational fishermen. Entry to the professional fishery is regulated by central government through the issue of individual fishing licences and vessel permits. These, together with specific regulations which may be introduced from time to time, determine who has the right to fish what, where, with what gears and during which periods of the year (Píriz, 1998). All other fishermen are regarded as fishing only during their free time; but a further distinction is made between those who fish purely for recreation - sports fishermen defined by their use of rod and line - and those who intend to supplement their household food consumption through fishing. These latter 'household' (*husbehovsfiskare*) fishermen may deploy a greater range of gears in public waters but will be restricted in the number of gears used, for example, a maximum of six nets with a combined maximum length of 180 m, six small traps, six pots (or 14 in the case of lobsters) and six longlines with a maximum of 100 hooks.

One of the problems associated with the presence of large numbers of unlicensed non-professional fishermen in Sweden is the inevitably sketchy information on their contributions to total fishing effort. The same is also true of the private water owners, whose catches are not always formally recorded. By contrast, licensed professional fishermen are required to submit logbook information detailing information on location, gears and catches on a daily or, in the case of inshore fishermen, a monthly basis.

6. EVALUATION OF THE INSHORE FISHERIES SECTOR

Throughout most of the past 50 years, Sweden's commercial inshore fisheries have been in decline as a source of employment. This has been particularly marked along the east coast especially in the aftermath of technological developments in herring fishing which effectively took control of these important seasonal fisheries out of the hands of artisanal coastal fishermen. Throughout Sweden the inshore sector has suffered a degree of neglect in terms of infrastructural investment in boats, harbour facilities, marketing and distribution and has been overtaken in importance by the recreational fisheries which are thought to account for some 75% of the total catch taken within the traditional 4 nm zone (Sandström, 1994; Andersson, 1998; Lagenfelt and Svedäng, 1999; Bengtsson 2000).

The inshore fisheries rely on a relatively narrow range of species. Again, this is more pronounced on the east coast and especially in the Gulf of Bothnia where the sub-Arctic regime is combined with low salinities characteristic of the semi-enclosed Baltic Sea. As a consequence the coastal fisheries are highly sensitive to fluctuations in stock size for the limited range of commercial species. Apart from the heavily exploited main species - cod, herring and salmon - relatively little is known of the underlying state of the stocks, their basic productivity and the current levels of fishing effort. Most are thought to be above safe biological limits and relatively underexploited, with the important exception of the eel which is suffering decline due to poor recruitment. Nonetheless, the range of commercial species is sufficient to allow the development of flexible patterns of exploitation involving the combination of different seasonal fisheries using a variety of fishing gears at relatively low levels of capital investment. The strength of the inshore sector lies in its ability to meet the demands of local and regional domestic markets for high quality, freshly caught fish - and there are ongoing attempts to enhance this particular attribute by the adoption of catching methods and pre-market storage which keep the fish in live condition.

On the other hand, the inshore sector has been struggling to compete with the more highly capitalised trawl fisheries targeting the same species (mainly cod and *nephrops*) in access to markets which do not sufficiently discriminate in terms of price for higher quality, fresh fish. Current management systems tend to privilege the offshore sector and so further marginalise the inshore fishermen. At the same time, local fishermen are finding it difficult to compete with affluent urban populations in the acquisition of coastal properties: the market price for coastal property has escalated dramatically in the more attractive areas of the archipelago which, combined with high land taxes, makes it impossible for fishermen to purchase or even hang on to shoreline properties with fishing rights. Access to private waters, especially in the Baltic, has become a major problem. In 1997, a survey of the east coast county of Kalmar revealed that only a third of all fishing waters less than 10 m deep were accessible to commercial fishermen as the result of complicated ownership patterns (Neuman and Píriz, 2000). Large areas of potentially rich fishing potential are thus grossly underused. Clearly, private ownership can act as a serious constraint on the optimal development of the inshore sector. Coastal fishermen are insufficiently well organised politically to protect their position against the thrust of central government policy and the burgeoning influence of recreational and wildlife conservation interests (see also Chapter 16).

Prospects for the future of the inshore sector remain uncertain. Much may depend on how far political concern for the welfare of the 'living archipelago' is translated into a balanced approach to management of the coastal zone and a realignment of the regulatory system to solve problems of access to coastal waters through the allocation of individual, local and regional fishing rights and the restoration and protection of areas for the spawning and recruitment of key commercial stocks. Failure to remedy weaknesses in the present system - including the organisation of the inshore fishermen themselves - will inevitably lead to a failure to reverse the negative trends in the recruitment of young people to the inshore sector.

One unresolved issue concerns the environmental status of the inshore waters, especially in the Baltic. There are many positive aspects to the current situation. The serious problems created by the pulp mills, mainly in the Gulf of Bothnia, have improved significantly in the 1990s. The national environmental monitoring programme indicates a continuing reduction of hazardous organic substances: DDT and PCB levels in herring have been declining at 5-10% per annum since the 1970s and mercury levels have been cut by a third since the early 1980s (Thoresson and Sandström, 1997). Concentrations of PCB in salmon and herring and mercury contamination of freshwater fish are now such that the National Food Administration's earlier recommendations concerning maximum levels of consumption no longer apply. Eutrophication remains a threat, altering the biodiversity of shallow coastal waters of significance for the recruitment of several commercial stocks which are also at risk from coastal construction works.

No environmental impact assessment has been made for Sweden's fisheries though it is widely held that coastal fishing causes comparatively little damage to marine habitats and ecosystems. Coastal fishing is considered basically sustainable as the main target species comprise many separate local stocks. General overfishing is thus an unlikely scenario. The main problem concerns the increased populations of predators, especially seals and cormorants. One consequence of the improvement in pollution levels in the Baltic has been the strong recovery of the grey seal population where fishermen are now seriously concerned about the damage to fish caught in the pound net fisheries for salmon and whitefish, where damage rates of 50-80% have been recorded (Lunneryd and Westerberg, 1997). Many seals also perish accidentally in static gears like the salmon trap net. As the environmental profile of the coastal fisheries depends crucially on the notion of coexistence with nature, the development of fishing patterns which both reduce the damage caused by seal and seabird populations and also protect them from accidental capture remains an important task for the coastal fisheries.

7. REFERENCES

Andersson, J. (1998) *Kustfisk och fiske vid Svenska Östersjökusten*, Fiskeriverket information 1: 1998, Fiskeriverket, Göteborg.

Anon (1998) *Fiskeriadministrationen i ett EU-perspektiv*, Översyn av Fiskeriadministration mm, Jordbruksdepartementet, SOU 1998:24, Stockholm.

Bengtsson, B. (ed.) (2000) *Fiske 2000: En undersökning om svenskarnas sport-och husbehovsfiske*, Fiskeriverket Informerar 1: 2000, Fiskeriverket, Göteborg.

Byron, R. (1993) Fishermen's organisations and the assertion of local interest in western Sweden, *North Atlantic Studies*, **3** (2), 32-37.

Gustavsson, T. (1998) *Swedish fishery in 1997. A summary of basic data*, Fiskeriverket information 7:1998, Fiskeriverket, Göteborg.

Hultkrantz, L., Hasselberg, Y. and Stigberg, D. (1997). *Fisk och fusk: Mål, medel och makt i fiskeripolitiken*, Regeringskansliet, Finansdepartementet, Stockholm.

Lagenfelt, I. and Svedäng, H. (1999) *Fisk och fiske i Västerhavets och Öresunds kustområden*, Fiskeriverket rapport 1999:7, Fiskeriverket, Göteborg.

Lunneryd, S.G. and Westerberg, H. (1997) *By-catch of, and gear damages by, grey seal (Halichoerus grypus) in Swedish waters*, ICES CM 1997/Q:11, Charlottenlund.

Neuman, E. and Píriz, L. (2000) *Svenskt småskaligt kustfiske - problem och möjligheter*, Fiskeriverket rapport 2000:2, Fiskeriverket, Göteborg.

Olsson, I. (1997) *Action plan on biological diversity*, Fiskeriverket information 5:1997, Fiskeriverket, Göteborg.

Píriz, L. (1998) A review of the Swedish coastal management in the light of coastal fishers' governing needs. Paper presented at the European Social Science Fisheries Network Workshop on *Inshore Fisheries Management*, Amsterdam, 24-26 September 1998.

Píriz, L. (2000) Dependence, modernisation and coastal fisheries in Sweden, in D. Symes (ed.), *Fisheries Dependent Regions*, Blackwell Science, Oxford, pp.116-131.

Sandström, O. (1994) *Kustfisk och fiske i Bottniska viken*, Kustrapport 1994: 1, Fiskeriverket, Öregrund.
Thoresson, G. and Sandström, O. (1997) *Resurs - och miljööversikt - Kustfisk och fiske*, Fiskeriverket information 3: 1997, Fiskeriverket, Göteborg.

Chapter 4

DENMARK

Eva Roth

1. INTRODUCTION

Denmark is a small country (population 5.3 million in 1999) but, in the context of the European Union, a large fishing nation. This is primarily due to the underlying geographical conditions for, apart from the Jutland peninsula, Denmark is essentially made up of a large number of islands. No one in Denmark can live more than 50 km from the sea. Compared to other Member States, Denmark takes a significant share of total EU landings (23% in 1998) but, because of the disproportionate volume of the catch intended for reduction to fish meal and oil, a much smaller share of total value (about 8% in 1997) and employment (less than 2%) arising from the Community's fisheries (Eurostat, 2000). Nonetheless, Denmark is by far the largest exporter of fish in the EU. By 1999 the total Danish catch had reduced from almost 2 million tonnes in 1995 to 1.3 million tonnes, with a turnover value of 3,100 million DKr.

The fishing sector has been in decline since the mid-1980s; landings have fallen and the real value of the catch has fallen even more rapidly. The fleet has also contracted both in terms of overall numbers and total tonnage. Between 1987 and 1993 a total of 815 vessels representing *circa* 40,000 GRT - or 30% of the total registered tonnage - was decommissioned. Over the same period, sea-going employment fell from around 7,000 full time fishermen to almost 5,000. Since the early 1990s these trends have continued though at a much reduced rate. This is seen in Table 4.1 which highlights an ongoing decrease in the number of Danish vessels, albeit over this period total tonnage has remained relatively constant.

Fisheries management in Denmark makes no formal distinction between inshore and offshore fisheries. The development of a two-tiered management

D. Symes and J. Phillipson (eds.), Inshore Fisheries Management, 61–78.

Table 4.1 The Danish fishing fleet 1996-1999

Year	Number of vessels	Tonnage (GRT)
1995	5,183	89,652
1996	4,831	86,825
1997	4,581	86,615
1998	4,374	86,448
1999	4,231	86,601

Source: Ministeriet for Foedevarer, Landbrug og Fiskeri, 2000

system differentiating between fishing activity within and outwith the 12 nm territorial limits makes little sense when faced by the diversity of natural conditions contained within the complex marine geography comprising not only the exposed open waters of the North Sea, Skagerrak and Baltic Sea but also the inner waters of the Kattegat, the Sound and the Belts. As a result, defining and delimiting inshore fisheries in Denmark constitutes a major problem. It may be attempted on the basis of geography, fishing patterns and behaviours or simply common sense, but official policy offers little guidance. Common sense suggests a definition of inshore waters in which all the fjords - and especially Limfjord - together with Øresund, dividing Denmark from Sweden, and the Belts which separate the islands of Sjælland (Seeland), Fyn and the Jutland peninsula are considered inshore waters. Opting for a 12 m vessel length definition of the inshore fleet may be open to criticism though significantly it represents the threshold below which disposable fishing capacity can be allocated to new vessels entering the commercial fisheries.

Commercial and recreational fisheries, together with aquaculture in the coastal zone, are regulated in a number of different ways. The Danish management system differentiates between full time commercial fishermen, part time fishermen, sparetime fishing for household consumption and sports fishing with rod and line; and these distinctions determine the extent and nature of access to fishing rights. The management system also imposes regulations in the form of vessel and gear restrictions in both marine and freshwater fisheries; this form of regulation has been greatly strengthened over the past ten years or so - first and foremost as a means of combating overcapitalisation in the fishing sector but also as a means of striking an appropriate balance between stock abundance, fishing effort and revenues from fishing. In effect, Danish fisheries management has traditionally been highly sensitive to local environmental conditions. As a result there is an abundance of local regulation, centrally imposed and enforced, including 29 orders regulating freshwater fisheries, 52 orders for coastal areas and fjords and a further 53 orders which include fisheries in the regulation of designated areas for wildlife conservation. This tradition is currently being maintained though the development of a fisheries plan for Denmark's largest area of inshore waters, the Limfjord (see below).

2. INSHORE FISHERIES IN DENMARK

2.1 Commercial capture fisheries

Danish landing statistics are based largely on the registration of sales notes from the first hand sales of fish made in both Danish and foreign ports; in addition all commercial fishing vessels over 10 m are required in law to provide logbook information detailing the species caught, the fishing areas and the place of landing. Together these data can provide a quite detailed picture of Denmark's fishing industry, though the exclusion of many commercial vessels under 10 m from the requirement to provide logbook returns, together with the non-recording of a significant portion of the recreational catch, renders the data liable to significant underrepresentation of the true volume of the total catch in general and the inshore catch in particular.

Nonetheless, Table 4.2 provides a valuable overview of Denmark's inshore fisheries in 1997 by combining information for all vessel landings from the fjords and from Øresund with details of landings by the registered under 12 m fleet in other areas. Although not offering a complete picture certain important features can be identified. First, the inshore fisheries account for over 100,000 tonnes of fish and shellfish valued at over 400 million DKr. Secondly, the pattern of fishing is dominated by the use of passive gears (fixed nets and traps) rather than mobile gears (trawls, dredges); two thirds of the first hand sale value of the inshore fisheries is attributable to passive gears. This situation is reversed in terms of the volume of landings where 80% of the tonnage landed comes from trawling or, in the case of the blue mussel, from dredging. Thirdly, the composition of the inshore catch overall is dominated by cod (38%) and other food fish species (25%) and this pattern is even more pronounced in the case of passive gear fishing where cod (48%) and flatfish (24%) together account for almost three quarters of the total value of the landings. Where mobile gears are used, the blue mussel (36%) is the single most important species. The miscellaneous category of 'other fish', which accounts for a significant share of the catches by vessels using passive-gears, includes a number of high value species, among them lumpfish caught essentially for their roe, anglerfish, eel and salmon.

Finally, while the use of mobile gears - and, presumably, a larger vessel size - clearly yields a much higher gross revenue (400,000 DKr per vessel),

Table 4.2 Denmark's inshore fisheries in 1997

	Mobile gears			Passive gears			Total		
	Vessels	Tonnes	DKr 000	Vessels	Tonnes	DKr 000	Tonnes	DKr 000	% value
Cod	261	4,594	39,719	1,164	11,985	114,489	16,579	154,208	38.4
Other gadoids	193	128	602	497	216	2,733	344	3,335	0.8
Plaice	257	700	9,571	906	1,740	25,157	2,440	34,728	8.7
Sole	197	55	4,306	701	310	25,732	365	30,038	7.5
Other flatfish	260	928	7,821	1,293	2,197	22,580	3,125	30,401	7.6
Herring, mackerel	86	4,420	6,326	370	1,650	4,794	6,070	11,120	2.7
Other finfish	247	109	2,228	1,399	3,082	66,272	3,191	68,500	17.1
Norway lobster	63	172	9,647	34	5	294	177	9,941	2.5
Shrimps	3	7	63	5	0	7	7	70	-
Blue Mussel	52	64,878	47,882	3	0	0	64,878	47,882	11.9
Other shellfish	104	165	2,065	576	273	4,351	438	6,416	1.6
Fish for reduction	20	4,823	4,473	5	116	84	4,939	4,557	1.1
Total	336	80,978	134,702	1,714	21,575	266,494	102,553	401,196	100

Source: Data provided by Fisheries Directorate of the Ministry of Food, Agriculture and Fisheries the unit value of landings persistently favours passive gear operations. Even when the landings of blue mussels, herring and mackerel and industrial fish intended for reduction are ignored, the prices paid for fish caught with passive gears are some 20% above those for fish caught with trawls. This presumably reflects the better quality of the fish landed from passive gear fisheries. Although not directly comparable with the landing data shown in Table 4.2, the economic returns provided by the Vessel Account Statistics (Fiskeri - og Regnskabsstatistik, 1997) suggest relatively poor performances by the inshore sector. Commercial net fishermen operating from vessels under 20 GRT - the typical Danish inshore fishermen - generate low labour incomes, with negative net outputs and negative rates of return at around 20%. Stratified for age, the lowest incomes occur among those under 35 and the highest incomes are recorded by those aged between 45 and 64.

When the overall pattern described in Table 4.2 is broken down regionally, the expected numerical dominance of the Baltic and Skagerrak/Kattegat areas, accounting for 33% and 28% of inshore landings respectively, is confirmed. Figures 4.1 and 4.2 also reveal some significant variations in the composition of the inshore catch and one striking anomaly in the case of Limfjord where the current pattern of fishing is totally dominated by dredging for blue mussels. Inshore fleets in the Skagerrak and Kattegat record the highest gross earnings, reflecting the greater importance of the high value flatfish landings when compared with the Baltic, for example, where the much larger volume of landings is offset by a greater reliance on the poorer quality Baltic cod.

2.2 Aquaculture

Aquaculture makes an important contribution to marine and freshwater fisheries in Denmark in terms of production for the market and the supply of brood stock to replenish partially depleted commercial and recreational fisheries, but for the past decade or so the aquaculture sector has been faced with the need for rapid redevelopment in order to meet the stringent regulations intended to safeguard the natural environment.

In Denmark the aquaculture sector can be divided into three more or less discrete branches (see Table 4.3): the production of rainbow trout mainly in freshwater locations which in 1997 totalled 31,957 tonnes with an estimated value of 482 million DKr (including fish, eggs for hatcheries and roe); the production of rainbow trout in sea cages; and the rearing of eel in intensive freshwater recirculation systems. In addition there is a small number of separate enterprises specialising in the production of juvenile turbot at three sites intended for the restocking of Danish waters and for export; plaice and

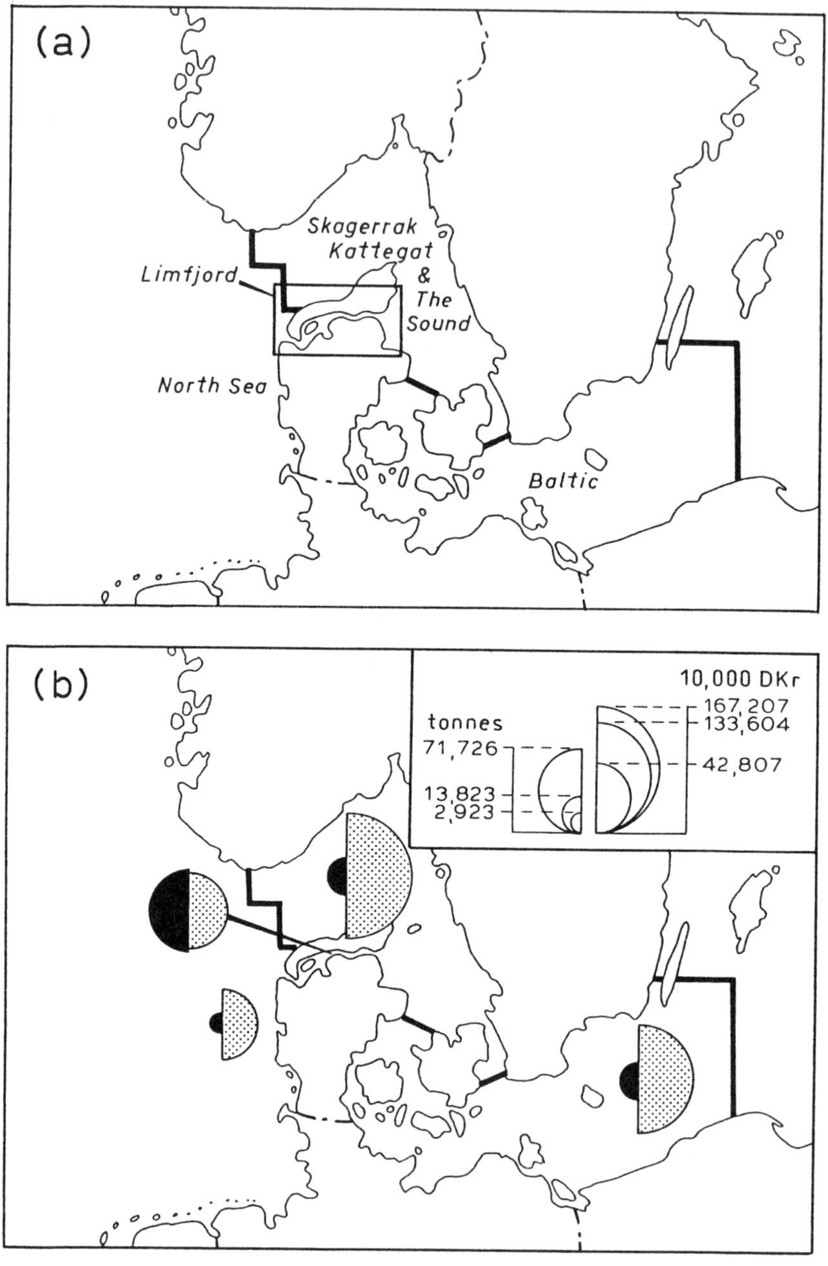

Figure 4.1 Denmark: Regional Pattern of Inshore Fisheries 1997: (a) Definition of Regions; (b) Landings by Volume and Value (Source: Ministry of Food, Agriculture and Fisheries, the Fisheries Directorate, Vessel and Sales Note Registers)

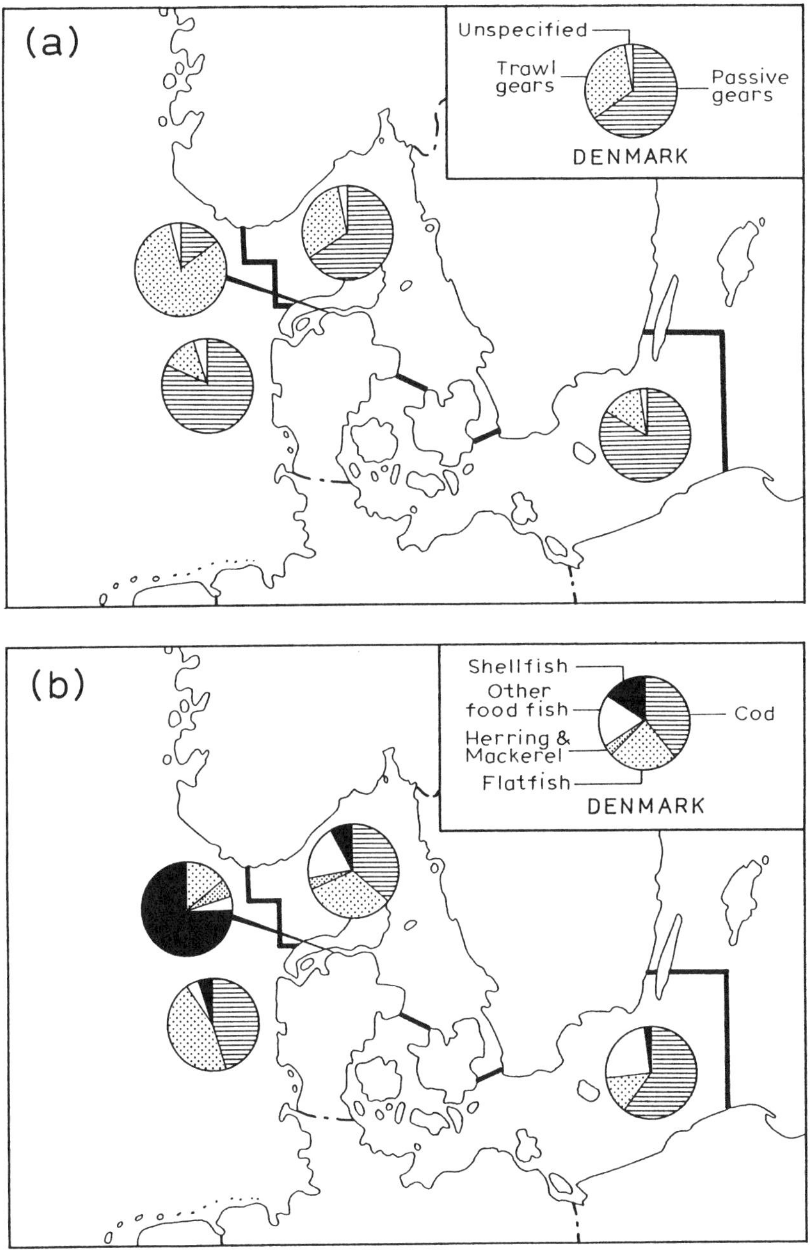

Figure 4.2 Denmark: Regional Pattern of Inshore Fisheries 1997 (continued): (a) Landings by Types of Gear; (b) Value by Species (Source: Ministry of Food, Agriculture and Fisheries, the Fisheries Directorate, Vessel and Sales Note Registers)

Table 4.3 Aquaculture production in 1998

	Trout		Fish eggs	Eels	Total
	Freshwater aquaculture	Mariculture	from mariculture		
Tonnes	32,607	7,053	400	2,500	42,560

Source: Frost *et al.* 2001; Olesen and Roth, 1999; Ministeriet for Foedevarer, Landbrug og Fiskeri, 2000

Table 4.4 Export of trout from freshwater aquaculture in 1996

	Volume	Value	
	Tonnes	Thousand DKr	%
Live trout	7,820	146,517	26
Fresh/chilled trout	2,480	55,891	10
Frozen trout	7,904	171,754	31
Smoked trout	2,094	143,938	26
Fillets, fresh	766	29,586	5
Fillets, frozen	250	9,683	2
Total	21,314	557,369	100

Source: Data supplied by Ministry of Food, Agriculture and Fisheries 1998 and Danish Trout Farmers 1998

flounder for restocking Danish waters only; salmon, trout, pike and whitefish for stocking freshwater lakes and rivers; and finally, oysters (1 plant), mussels at two sites in Mariager fjord, and freshwater crayfish.

The balance of aquaculture in Denmark strongly favours freshwater production systems with 433 plants, compared with only 28 based on saltwater locations. Moreover, freshwater aquaculture is concentrated in relatively few hands with roughly a fifth of all producers responsible for almost two thirds of the total sales of farmed trout. Much of the production is intended for the export market, with a focus on high quality, value added production (see Table 4.4). Some commercial trout farmers also cooperate with private angling clubs in the restocking of local sports fishing waters. The farming of eels, which depends on the supply of wild caught glass eels, is contained within a total of 30 units (1996), many of which are owned and operated in conjunction with commercial agricultural enterprises. Again export markets are the principal targets and again the emphasis has been on quality rather than quantity of production so as to compete effectively with French, Spanish and Italian producers.

Direct employment in aquaculture falls somewhat short of a thousand with 613 employed full time in the freshwater sector, 200 in the saltwater based plants - though this figure may be doubled in October and November with the addition of seasonal workers - and 55 on the eel farms.

Environmental regulations have insisted on changes in the production technologies associated with all forms of aquaculture in order to meet national and EU water quality standards, *inter alia*. Probably the most stringent demands are imposed on freshwater production systems for rainbow trout, where there is a need to ensure that the discharged waters are free from pollutants potentially associated with intensively farmed fish. On the coast, land based saltwater plants use pumped seawater, which is returned to the sea after treatment, but such systems are expensive to install and operate especially at a time of relatively low market prices for sea trout. Because of the risk of pockets of oxygen depleted water, marine cultivation in sea cages is limited to a six or seven month production year. Experimental systems are being developed involving the installation of oxygenation systems under the cages which will both optimise the oxygen conditions and help achieve improved feed conversion rates.

The legal framework for establishing aquaculture units is very restrictive. Basically the county councils - as the planning authorities - are obliged to take into consideration a whole raft of national regulations relating to the protection of the environment, the maintenance of public water supplies and land drainage systems when making their decisions. The setting up of a sea cage operation requires the permission of both the Ministry of Food, Agriculture and Fisheries and the Ministry of the Environment, together with the county authority. The issuing of new permits was suspended from 1986 to 1991 and all existing permits expired in 1992. At present there are some 28 plants in operation on time limited licences.

All mariculture and freshwater aquaculture licences are due for renewal in 2005 and it is unlikely that any additional permits will be granted in the interim. The licensing of plants also involves the attaching of strict conditions relating to the amount and quality of feed, the ratio of feed to each kilogramme of fish produced and permitted levels of phosphorus (P) and nitrogen (N) discharges, set against national targets for the maximum nutrient load of seawater. It is this last consideration that led to the moratorium on the issuing of permits for new seawater based fish farming units, as well as limiting the development of freshwater aquaculture.

The regulation of eel production is governed by significantly different objectives. Unlike either rainbow or sea trout production, eel farms are not subject to restrictions over the use of feed and the waste waters are most frequently discharged onto agricultural land rather than directly into water courses. Moreover, a number of innovative producers have adopted water purification technologies enabling the reduction of P and N content to very low levels. The aim of the industry is to develop a public environmental certification system which will help it qualify for market-led incentives to provide ecologically friendly products.

3. RECREATIONAL FISHERIES

As indicated in the introduction, fisheries management in Denmark distinguishes two separate categories of recreational fishermen. First, there are those who fish in their spare time with a restricted number of passive gears and who take fish for their own household consumption or barter their catches locally, but whose catches are nominally included in official landing statistics. In 1997 a total of 33,559 licences were issued to such fishermen at an individual cost of 250 DKr per annum. Secondly, there are the sports fishermen restricted to the use of rod and line in either fresh or salt water locations; if between the ages of 18 and 66, they too are required to have a licence, paying either 100 DKr for an annual licence, 75 DKr for a weekly permit or 25 DKr for a day's fishing. It is estimated that in 1996 around 425,000 - or roughly one in eight Danes aged 18-66 - actually engaged in sports fishing; of these women accounted for 15%. A total of around five million days were spent fishing by these sports fishers. This segmentation of Danish fishermen into commercial and sparetime, introduced into Danish fisheries management in 1978, broke with the established tradition of free access to the fisheries. It was intended to reserve access rights in the first instance for commercial fisherman who must earn their living from fishing.

Among the group of sparetime or household fishermen, three different types can be identified (Hansen, 1996): former commercial fishermen who have retired from full time fishing and taken up jobs on land; those who are looking to gain experience and income from fishing with a view to qualifying for a commercial licence; and the traditional household fishermen who regard fishing as neither a hobby nor a regular revenue earning activity but rather as a means of supporting a traditional lifestyle and enjoying membership of a social group based around the fishing harbour. For this last type, the costs of fishing should be recouped either through incomes in cash or in kind or through savings in household expenditure on food. However, in 2000 a moratorium on all trade in fish caught in the recreational fishery was introduced.

The most restrictive aspect of management of the sparetime fishermen concerns the limitations on the numbers of gears that may be used. Currently these restrictions impose an overall limit of six gears - either six lines with up to 100 hooks or six nets or six traps. Further restrictions are also placed on the type and location of the nets, which may not be set closer than 100 m from the low water mark nor at the entrances to rivers in order to protect migrating fish, principally trout and salmon.

A survey of angling undertaken in 1997 enables a greater insight into the distribution of angling activity and the effectiveness of the licensing system (Bohn and Roth, 1997). Of the 396 anglers included in the survey 73% had fished in coastal waters, 30% in lakes and 25% in streams, together with 27% in 'put and take' waters not covered by licence restrictions. Strong regional differences were noted with a significantly higher proportion taking part in sea angling in the eastern part of Denmark (83%) compared to the west (63%). The general emphasis on sea angling throughout Denmark reflects the simple fact that whereas fishing rights in inland lakes and rivers are normally tied to riparian ownership, sea angling is open to all who hold an appropriate licence. Questions relating to licence holding left some margin of doubt as to how effective the licensing system is in actually defining the true number of anglers. Only a small fraction of those included in the survey were members of an angling association, despite the fact that such membership may increase the opportunities for fishing freshwater lakes and rivers through rental agreements between the association and the landowner. But this finding was significant for another reason, namely that non-governmental organisations - including angling associations - are permitted in law to hold seats on the saltwater and freshwater advisory committees and to be represented at hearings on environmental matters or appeals relating to administrative decisions taken under the *Environmental Protection Act, 1974*.

Inevitably there are disputes between sports and sparetime fishermen over the right to fish for certain high value migratory fish. One of the main arguments deployed by the sports fishermen is that their contribution to restocking programmes through the licence fee system is much greater. Indeed, some angling clubs have established their own hatcheries - often in cooperation with the local or regional authorities - in order to assist the conservation of stocks and enhance the local attraction for sports tourism. As a rule of thumb, three out of every four trout caught in Danish waters can be attributed to restocking programmes. Moreover, income from the sale of licences to both sports and sparetime fishermen is allocated to 'fish care' projects, including research, river improvement and restocking, organised through the Danish Institute of Fisheries Research with a total budget in 1997 of 31.7 million DKr.

4. FISHERIES MANAGEMENT

4.1 Legislative basis and organisational structures

Participation in the commercial fisheries requires both the registration of the individual fisherman and the registration and licensing of the fishing vessel. The conditions for registration as a commercial fisherman require Danish citizenship or permanent residence in Denmark for a period of two years prior to registration and employment in the commercial fisheries for the previous 12 months, during which time fishing must have contributed at least 35% of gross personal income. Only registered commercial fishermen are permitted to employ non-registered fishermen among their crew, where the only condition for employment is a three week period of safety training. In order to use a vessel for commercial fishing it must be registered as a commercial fishing vessel and at least two thirds of the vessel must be owned by persons eligible to fish on a commercial basis. At present the registration of new vessels is restricted by the need to ensure that the total disposal capacity of the fleet agreed under the EU's Multi-Annual Guidance Programme is not exceeded. The purpose of these entry qualifications is to restrict commercial fishing to those who derive their primary income from the fisheries, while at the same time safeguarding the industry against the threat of overcapitalisation.

Part time fishermen - those relying on commercial fisheries for less than 35% of their gross personal income - are recognised in Danish law as a separate category under broadly the same conditions except that they are not allowed to employ paid (or even unpaid) labour to assist in their fishing activities. Additionally, sparetime fishermen may apply for qualification as part time commercial fishermen within a specific geographical area.

As with all Member States in the EU, the basic rules for the management of marine capture fisheries are laid down in the Common Fisheries Policy and apply both outwith and within the 12 nm limits. As a consequence several species of relevance to the inshore fisheries - most notably cod, sole and plaice which account for almost two thirds of total landings by value from the inshore sector - are subject to quota management and to other restrictions relating to minimum landing sizes and bycatches. In addition the coastal fishery for salmon in the Baltic is regulated through quotas recommended by the International Baltic Sea Fisheries Commission. The implementation of these and other EU regulations referring to structural and marketing measures is the responsibility of the Ministry of Food, Agriculture and Fisheries, while the *Law on the Regulation of Fisheries 1993* lays down the objectives of national policy together with the instruments available for the implementation of EU and domestic legislation. In addition, two separate pieces of legislation - the *Law on Freshwater Fisheries 1992* and the *Law on Saltwater Fisheries 1993* - define more precisely the competence of the Ministry in terms of fisheries management, under which various forms of regulation can be initiated. Danish fishing law is very extensive as indicated by the fact that there are no fewer than 334 national and local legal orders pertaining to freshwater and marine fisheries. Many of these apply particularly to coastal waters.

Although fisheries management is highly centralised and relies strongly on a top:down mode of implementation, policy formulation is largely conducted through a process of negotiation with representatives of the fishing industry. Consultation with the industry has been greatly simplified following the merger in 1994 of the two 'national' fishermen's organisations - one representing large boat interests located mainly in Jutland and the second reflecting the small boat sector. The single organisation, *Danmark's Fiskeforening* (DF), representing around 80% of the registered commercial fishermen through the federation of some 86 local associations, is much better placed to negotiate with the central authorities from a position of strength. The principal arena for such negotiations is the ten member Regulatory Advisory Board, where DF occupy two seats alongside the General Workers Union, together with single member representation from the producers' organisations (POs), the processing industry, fish meal producers, the canning industry, the Danish Institute for Fisheries Research and the Ministry. Individual sub-committees deal with specific fisheries and their memberships vary accordingly. The POs, whose development has been characterised by fission rather than fusion, have little or no direct involvement in fisheries management; they are concerned primarily with the implementation of EU marketing policies. The participation of the industry in decision making is further exemplified by the provisions of the Freshwater and Saltwater Acts for the establishment of advisory committees whose membership is derived largely from the appropriate fishermen's organisations. In the case of freshwater fisheries there is scope for setting up local committees to recommend to the Ministry the adoption of local regulatory measures. Similarly, under the Saltwater Fisheries Act, an advisory committee involving the representation of local fishing interests has been formed for the management of the Danish sector of the Wadden Sea.

4.2 Limfjord: a case study of local inshore fisheries management

An example of the local management of inshore fisheries is provided by the Fisheries Development Plan for Limfjord which is being put together by a consortium of interests including the Fisheries and Environment Ministries and the three coastal counties of Nordjyllands, Viborg and Ringkoebing, with significant inputs from local commercial and recreational fishing interests. The Plan will need to reflect the growing concern for the integration of fisheries and environmental management and resolve the competing interests of several different groups.

Limfjord occupies a unique location in Denmark's inshore fisheries. Classed as a fjord, it is in fact a channel open at one end to the Kattegat and

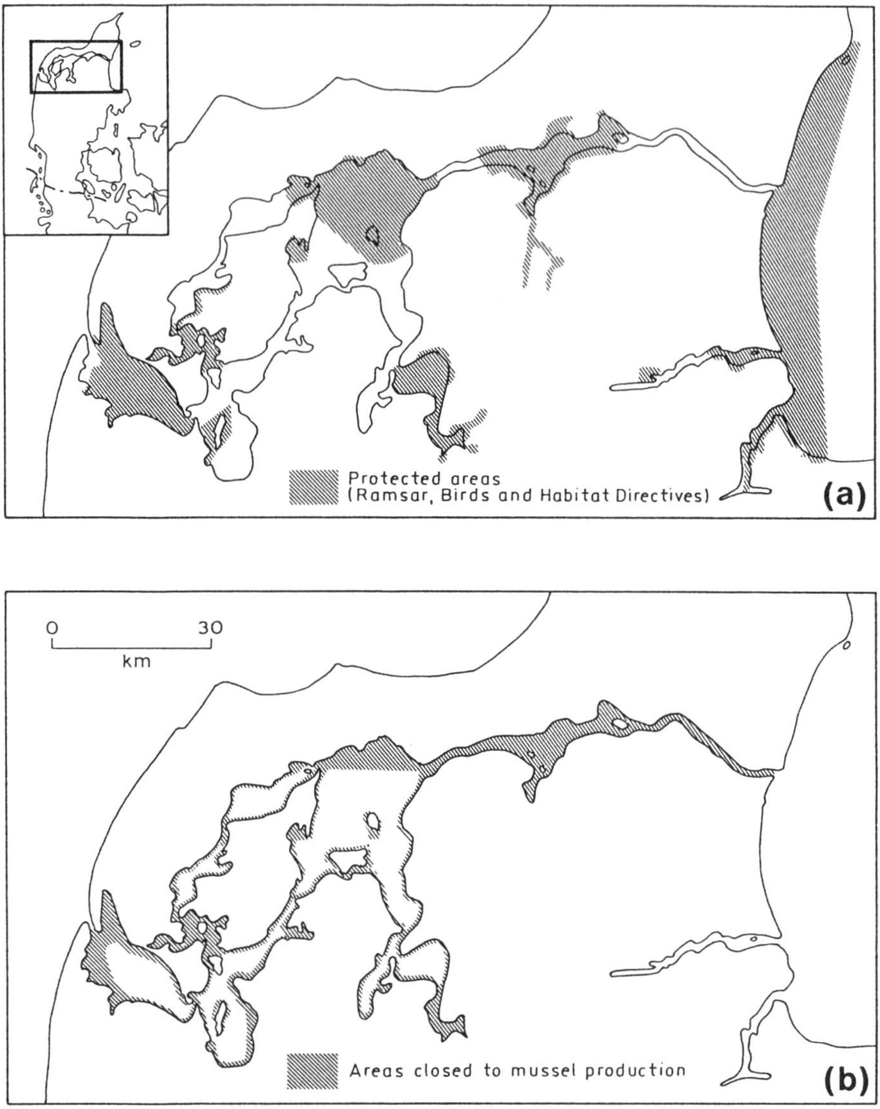

Figure 4.3 Limfjord: (a) Protected Areas (Source: Anon 1999) (b) Areas Closed to Mussel Production (Source: Data supplied by Ministry for Food, Agriculture and Fisheries, 1997)

at the other, via the Thyboroen Canal, to the North Sea (see Figure 4.3). Its hydrographic and ecological character is that of a shallow, eutrophic fjord, measuring 150 km in length and containing a surface area of 1500 km^2 with a mean depth of 7 m and varying in salinity from 1.5 to 2.8%. Until the early 1950s its fisheries were based on cod, plaice, herring and eel caught with trawls, set nets and pound nets. In the 1950s an industrial trawl fishery for

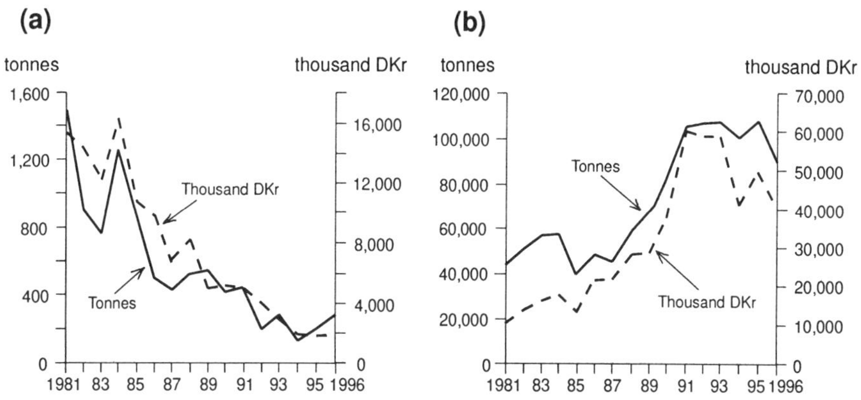

Figure 4.4 Limfjord Fisheries 1981-1996: (a) Landings of Finfish; (b) Production of Blue Mussels (Source: Data supplied by Ministry for Food, Agriculture and Fisheries, 1997)

sprat and small herring was introduced; it peaked in the early 1970s and has since dwindled to a relatively low level. Taking its place in the 1970s and gradually increasing in response to strengthening market demand was the blue mussel (*Mytilus edulis*): landings doubled between 1985 and the early 1990s when they exceeded 100,000 tonnes gross weight (see Figure 4.4). Today there are 51 small boats under 8 GRT licensed to dredge for mussels in the fjord. Overall employment in the fishing industry has declined from *circa* 500 in 1980 to less than 200 at the end of the 1990s.

The changing pattern of fisheries in Limfjord can only be explained by reference to the changing environmental conditions and to the increased nutrient load created by urban, industrial and especially agricultural waste water discharges. The heavy application of N and P based fertilisers in particular has led to a doubling of phytoplankton production causing both a reduction in the traditional finfish stocks and the explosive growth of blue mussel production. In its turn, the dredging of mussels impacts on the substrate and the ecosystem sparking off a clash of interests between the mussel men on the one hand and the recreational fishermen and conservation lobbies on the other, provoking demands for the stricter control or even the complete abandonment of the mussel fishery. This latter option would be a very drastic remedy for an industry which today accounts for 85% of Limfjord's fishing revenue and provides employment in fishing and processing for around 300 persons (Anon, 1999).

Although mussel production is by far the most important source of employment and income, the aim of the Limfjord plan is to provide a broader spectrum of commercial fisheries within the context of a 'cleaner' aquatic environment and a more diverse marine ecosystem which will meet the demands of a wide range of interests including recreational fisheries, nature conservation, water sports *inter alia*. The basic problem is how to enhance the opportunities for these non-commercial fishing interests without seriously undermining an important local economic asset. Only a quarter of a century ago, inshore fisheries like those in Limfjord were largely unregulated (Hansen, 1995). Today the mussel fisheries are already highly regulated. In addition to bans on fishing covering over a third of the total surface area, the mussel fishery is subject to closures for up to ten weeks during the summer; daily and weekly landing limits are imposed; and there are restrictions on the number and size of the fishing vessels and on the size specifications of the dredges. Over the long term, however, the aim must be to reduce the scale and impacts of the mussel fishery. This is most likely to be achieved by a reduction in the number of licensed vessels, a redefinition of the areas where dredging is permitted, the development of less damaging fishing practices and the possibility of mussel cultivation in certain areas of the fjord (Anon, 1999). So far there is little evidence that improvements in water quality have adversely affected the reproduction of the natural mussel beds. It remains to be seen, however, how far improvements in environmental quality will help to re-establish commercially viable stocks for other species.

5. EVALUATION

The foregoing analysis has identified a diversity of activities which coexist - not always comfortably - within the concept of inshore fisheries. Each has a legitimate role to play in the present and future development of marine resources. Among the more overt forms of commercial exploitation, the capture fisheries have already demonstrated their ability to supply the markets with high quality products, involving relatively low levels of investment and with comparatively little risk to the environment, except in the case of mussel production. Systems of production remain reasonably flexible and are not necessarily dependent on quota limited species. But problems arise from the low incomes related to the limited number of days fishing and from the high average age of a fleet sorely in need of capital renewal. The unattractive economic prospects discourage the entry of young people and this raises questions about the social sustainability of the sector. Two opportunities could help to reverse these trends: subsidies directed at the renewal of the inshore fleet and the establishment of young fishermen; and the development of a more discriminating market for certificated quality produce achieved through ecologically sustainable fishing practice. But there are also risks to the recovery or even survival of small boat fisheries from the persistence of low quayside prices and a further weakening of their competitive position and from the tightening of environmental policies which further restrict inshore fishing activity.

In marked contrast, aquaculture's principal strengths - apart from the favourable natural conditions for its development - lie in the innovative business environment which helps to promote high levels of efficiency in production and marketing. Its main weaknesses - and threats to its future development and expansion - relate to its negative public perception, the strength of opposition from other user groups and the stringent environmental controls which currently involve a moratorium on the licensing of new production sites in coastal and inland locations.

Despite its inclusion in the broad category of 'recreational fisheries', sparetime or household fishing has much more in common with commercial fishing than sports fishing. Although consumption replaces income as the key economic benefit, household fishing may be seen as part of the life cycle of commercial fishing, providing a means of entry to and retirement from full time commercial activity. It remains an integral part of the social ecology of inshore fisheries helping to maintain the social networks and values associated with low intensity, small boat fisheries, especially, in smaller, remoter communities. However, it faces continuing opposition from better organised and more influential commercial and sports fishing lobbies and is thus subject to continuing pressure to confine still further the opportunities for household fishing. It stands accused, for example, of creating black or grey markets for fish which undermine prices on the official markets, and of illegal fishing for migratory salmon and trout - target species for sports fishermen.

The very nature of sea angling contrasts sharply with the preceding categories of inshore fishing. Non-commercial in orientation (except where associated with tourist development) and much more closely identified with urban society, it forms an important element of the 'fourth wave' of recreation. It is growing in popularity and expansionist in outlook, well organised and thus able to exert a strong influence on fisheries and environmental politics, and also closely identified - in its own image, at least - with healthy lifestyles and an appreciation of nature. Notwithstanding its political influence, sports fishing lacks a clearly defined long term strategy, preferring instead to solve conflicts on a case by case basis. As victim of its own popularity, sports fishing faces problems of 'overcrowding' on the accessible inland waters thus lowering the quality of the recreational experience. Freshwater angling may be forced to pay higher prices for access to privately owned inland fishing opportunities. Sea angling may not have to contend with these same problems of access but it does face strong opposition from other user groups in inshore waters.

Management of inshore waters is essential to guarantee that each of these different interest groups is able to fulfil its legitimate economic and/or social role without detracting from opportunities of the other fishing or non-fishing activities. Future strategies are likely to focus on balancing the allocation of fishing rights and avoiding gear conflicts through the spatial separation of specific types of activity. Management is a necessary but not a sufficient condition for the sustainability of inshore fisheries: individual skill, business acumen and market performance will always be influential in determining the economic viability of the individual commercial operator.

6. REFERENCES

Anon (1999) *Plan for Fremtidens Fiskeri i Limfjorden*, Ministry for Food, Agriculture and Fisheries, Copenhagen.

Bohn, J. and Roth, E. (1997) Survey on angling in Denmark, 1997, in A.-L. Toivonen and P. Tuunainen (eds.), *Socio-economics of Recreational Fishery, TemaNord* 1997: 604, Nordic Council of Ministers, Copenhagen, pp. 79-89.

Eurostat (2000) *Fisheries Yearbook 2000*, Eurostat, Luxemburg.

Fiskeri - og Regnskabsstatistik (1997) *Statens Jordbrugs - og Fiskeriøkonomisk Institut*, Serie F No. 3, Ministry for Food, Agriculture and Fisheries, Copenhagen.

Frost, H., Jensen, C.L., Nielsen, M., Vestergaard, N. and Pedersen, M.I. (2001) *A socio-economic cost benefit analysis of the use of glass eel*, Ministeriet for Foedevarer, Landbrug og Fiskeri, Statens Jordbrugs - og Fiskerioekonomiske Institut, Rapport nr. 118, Copenhagen.

Hansen, K.M. (1995) The socio-economic aspects of the conflicting interests in a Danish coastal fishery. Paper presented at the *XVI European Congress on Rural Sociology*, Prague, July 31 - August 4, 1995.

Hansen, K.M. (1996) *Kampen om Limfjorden: livsformer, miljøværdier og reguleringsformer*, DFU Report No. 7, Danish Institute of Fisheries Research, Charlottenlund.

Ministeriet for Foedevarer, Landbrug og Fiskeri (2000) *Fiskeristatistist Aarbog 1999*, Fiskeridirektoratet, Koebenhavn.

Olesen, N.K. and Roth, E. (1999) *Det danske dambrugserhverv - en strukturanalyse*, IME Report 1/99, University of Southern Denmark.

Chapter 5

THE NETHERLANDS

Rob van Ginkel

1. INTRODUCTION

The Dutch fishing industry is one of the smallest in the EU. Including the supply sector and retail trade, it employs *circa* 15,000 or less than 0.4% of total employment. In 1999 the fleet comprised 399 beam trawlers (cutters), 16 pelagic freezer trawlers and 77 mussel boats (Table 5.1) with a combined manpower of some 2,560 persons. Since 1993, the year of the influential policy document *Vissen naar evenwicht*, the number of vessels in the Dutch fleet has declined by some 13%. In terms of economic output, mussel cultivation, shrimping, cockle fishing and oyster farming dominate the inshore fisheries accounting for 24% of the total value of landings in 1999.

Table 5.1 Size structure of the Dutch fishing fleet (number of vessels)

	1993	1994	1995	1996	1997	1998	1999	% change 1993-99
Cutter (1-300 hp)	244	247	239	234	230	224	224	-8
Cutters (301-2000 hp)	156	146	144	136	120	120	114	-27
Cutters (2001 hp and over)	74	71	69	67	66	63	61	-18
Trawlers	12	12	12	14	14	15	16	+33
Mussel boats	77	77	76	75	75	76	77	+/-0
Total	563	553	540	526	505	498	492	-13

Source: Visserij in cijfers, 1999

D. Symes and J. Phillipson (eds.), Inshore Fisheries Management, 79–96.

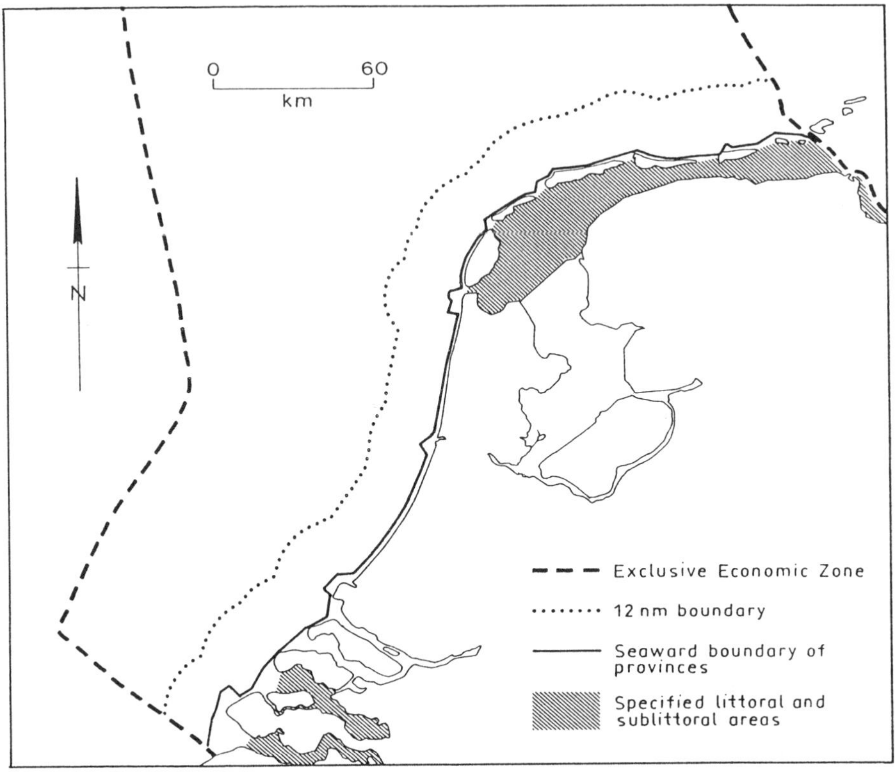

Figure 5.1 The Netherlands: Fishing Limits and Coastal Fishing Waters

Following EU directives, the 12 nm limits would seem to make a useful distinction between inshore and offshore fisheries. Vessels not exceeding 300 hp engine capacity are allowed to fish within these limits and in 1999, 224 such vessels had access. However, policy documents referring to *Kustvisserij* (coastal fishing) refer to fisheries in specific littoral and sublittoral areas specified in an amendment to the 1963 Fishery Act. They include: the Wadden Sea and the Dutch part of the Eems-Dollard estuary; the Western Scheldt, the Eastern Scheldt and its tributaries in the province of Zeeland; and the Voordelta and the Zeeland islands and peninsulas (Figure 5.1). These areas are extremely significant as spawning and nursery grounds.

Unlike offshore fisheries, the coastal fisheries are mainly subject to national policy the goals of which are: to regulate fisheries in relation to other marine functions, including the most important ecological function; to regulate fisheries *vis-à-vis* each other, and the leasing of shellfish farming plots. Recently, a system of devolved management has been introduced giving responsibility to the Dutch fisheries sector to develop a balance between fishing activity and nature through self-management.

2. INSHORE FISHERIES: THE MANAGEMENT APPROACH

There is growing pressure for the integration of fisheries and marine environmental management. Since 1993 the policy document *Vissen naar evenwicht* (Fishing for Balance) has sought to create a responsible way of fishing in which economic and ecological interests are balanced so as to achieve both economic and ecological sustainability (Vissen, 1993). The state aims to further fisheries interests through regulation, consultation with the industry and the maintenance of quality. Fishermen must take account of other functions of the seascape, especially its natural value. Fisheries policy is also intertwined with government policies regarding water management, nature conservation and planning, which makes for a complex institutional framework, especially for inshore fisheries.

Based on the *Nature Conservation Act 1974* the greater part of the Wadden Sea (150,000 ha) was designated a publicly owned nature reserve in 1981. In 1993 most of the remaining area was incorporated in the designation, which now covers 85% of the Dutch Wadden Sea. Several non-binding regional plans refer to the area but the key policy document is the *Planologische Kernbeslissing Waddenzee* (1981) (Physical Plan for the Wadden Sea, also known as the Wadden Sea Memorandum), defining the overall objectives for conservation, management and use of the Wadden Sea which are binding on all state, regional and local authorities. A number of boards and committees advise the politicians at different levels concerning proposed policies and plans (Steins, 1997). In addition, the Wadden Sea is subject to international agreements. In 1982, a Joint Declaration on the Protection of the Wadden Sea was agreed between Denmark, Germany and the Netherlands outlining their intention to coordinate future activities for the protection of the area. Trilateral conferences are held every three years. The Wadden Sea is also designated an EU Special Protection Area (1991), a Ramsar site (1984) and a UNESCO Man and Biosphere Area.

The Dutch state opts for a 'principle of interwovenness' in its fisheries policy for the Wadden Sea and Zeeland waters, meaning that exploitation is allowed only if fisheries and nature can coexist. Otherwise the functions are strictly separated. In the 1993 policy statement new forms of cooperation were encouraged. The state's aim is to enhance responsibility and social control through self-management. The state has considerable leeway in managing its coastal fisheries, principally through a system of licensing. Since 1993, inshore fishermen operate under a management plan (*beheersplan*). A co-management scheme for the mussel and cockle fisheries has been developed by two working groups, involving representatives from the mussel and cockle sectors, environmental organisations and the government. Producers' organisations (POs) have been established on a basis of voluntary membership: their prime goal is to achieve rational exploitation of the resources and improved marketing conditions. Fishermen's organisations contribute to the development, implementation and enforcement of management measures helping to increase user group support for the management regime including measures to protect nature (Keus, 1997).

The two main goals of the management plan are to limit exploitation of the marine domain and prevent food shortages for seabirds. In years when cockle stocks are low, 60% of a specified minimum level of the cockle and mussel supply is reserved for seabird consumption. When stocks fall below this level temporary bans may be introduced. Also in 1993 it was decided to close more than a quarter of all areas exposed at low tide, including the intersecting gullies, to mussel seed, cockle and other fisheries. The same applies to a large portion of the Eastern Scheldt while cockle and mussel seed fishing in the Western Scheldt and the Voordelta is subject to restriction. To protect the ecosystem of the Voordelta, the government also seeks to reduce shellfish dredging with fishermen agreeing to leave old mussel banks and eelgrass plots untouched. Annual total allowable catches (TACs) are set for the mussel and cockle sectors. In the mussel industry, the TAC (amounting to 65 million kg of mussel seed as of 1998) is divided into individual quotas. When the total TAC is exhausted the fishery is closed. The TAC for cockles is 10 million kg of cockle meat.

These measures are implemented through annual fishing plans drafted by the two POs and are binding on their members. The plans detail the area and time for fishing, gear restrictions, TACs, capacity reductions, control and arbitration procedures. All fishing vessels must be equipped with a 'black box' which records the boat's position and activity so that all fishing activities can be monitored. This devolved management system also implies that state supervision can be maintained at existing levels without need to intensify controls. Sanctions for breaching the rules are severe: mussel seed fishermen can be fined up to Dfl 20,000 and cockle fishermen can lose their permits. In March 1998 a review of the first phase of the new policy concluded that the management and the fishing plans developed by the fishermen's own organisations had contributed to the realisation of the fisheries policy (Structuurnota, 1998). On the basis of this it was decided to continue the policy in a second phase from 1998 to 2003 (Beheersplan Schelpdiervisserij, 2000). To contribute to the recovery of natural mussel banks, the cockle and mussel fishermen have decided voluntarily to close an additional 5% and 10%, respectively, of their fishing grounds.

There are, however, signs that the fragile relationships between the key parties in the co-management process are coming under considerable pressure. Development of the second management plan (1999-2003) is proving much more difficult (Steins, 2000) partly for technical reasons - the short time span allowed for negotiation following the evaluation exercise and the absence in the early stages of a carefully selected steering group to supervise the process - but also because of an erosion of trust between the fishing industry and the conservation organisations. This grew out of the latter's public denunciation of the mechanised cockle fishery in 1997, supported by some fisheries scientists and seen by the industry as reneging on their original commitment to co-management. The fisheries department was put in a difficult position, needing to act as facilitator in the negotiations but also expected to uphold the government's declared policy. Certainly the potential for consensus building has been seriously impaired by these developments.

3. THE FISHERIES

In terms of landing value and employment, molluscan and crustacean fishing and farming are the most important coastal fisheries (Table 5.2). Blue mussels (*Mytilus edulis*) and European flat oysters (*Ostrea edulis*) are cultivated, mussel seed, shrimps (*Crangon crangon*) and cockles (*Cardium edule*) are fished.

Table 5.2 Value of shellfish and crustacean landings (million Dfl)

	1990	1991	1992	1993	1994	1995	1996	1997	1998	1999
Mussels	102	91	92	88	102	120	107	123	101	125
Shrimps	46	39	44	54	43	68	58	42	53	85
Cockles	29	8	29	27	24	18	7	10	60	50
Flat oysters	4	5	8	8	7	4	6	6	5	8
Total	181	143	173	177	176	210	178	181	219	268

Source: Dutch Fish Product Board

3.1 Mussel farming and fishing

Mussel farming in the Netherlands is a semi-culture. For some weeks in spring and autumn, for three days a week, mussel farmers are allowed to catch young seed mussels. The Ministry of Agriculture, Nature Management and Fisheries (MANMF) which has jurisdiction over the shellfish grounds determines the opening and closing dates of this short season and the locations where dredging is permitted. The Fish Product Board (*Productschap Vis*), an organisation representing the fishing industry as a whole, together with representatives of the Mussel Advisory Committee drawn from the harvesters, merchants and processors, determine quality standards and set minimum prices.

Seed fishing forms the basis of cultivation and permits are granted only to mussel farming firms; a total of 88 permits are issued and since 1991 each firm is allowed to use one vessel to fish seed mussels. The musselmen plant the seed and young mussels on plots rented for a period of three years from MANMF. A total of 3,500 ha (divided into 460 plots) is available for cultivation in the Wadden Sea and 2,000 ha (380 plots) in Zeeland waters. Each firm rents a number of plots in each of the two areas at rentals based on acreage and the share of the total value of mussel production. Access and use rights are exclusive.

The natural beds are usually - though not always - productive enough to permit seed fishing from year to year. It is of paramount importance that fishermen catch a sufficient amount of seed to relay their plots. The mussels mature within two years by which time they are dredged and transported to the national auction at Yerseke, the country's foremost shellfishing community with 45 of the 84 mainly family owned mussel firms. No single firm has more than a 7% share of the total annual production. Processing and marketing of mussels is almost entirely concentrated in Yerseke where the top four companies have a 50% share of mussel sales. Throughout the 1990s, up to and including the 1999/2000 season, the auction paid an average of Dfl 106 million to the producers, roughly 10% of the value of all fish landings in the Netherlands. The average net profit was almost Dfl 450,000 per firm. Some 80% of the mussels are for export, mainly to Belgium and France.

Sharp fluctuations in production characterise the industry due usually to the scarcity of seed mussels on the natural beds but also to harsh winters, ice drift and storms which cause low recruitment. High prices tend to compensate for lower production. However the continuation of mussel seed fishing despite declining productivity has led to misuse and the depletion of the natural mussel beds.

A dedicated PO was established in 1988. Only three cultivators opted out, refusing to sign the fishing plan. The PO has since petitioned for an *extension de régime*, requiring non-members to comply with the fishing plans, but so far the Ministry has refused. The PO and the Fish Board negotiate with the Ministry over replacement of plots which have become unproductive due to siltation, land reclamation, hydraulic works or other cause. The state's aim is to optimise the use of existing plots rather than entertain any expansion. With demand far exceeding supply, there is scope for the industry to expand and remain sustainable but at the cost of diversity. A further increase in output would require more space and so reduce opportunities for other uses - or non-use - of the marine environment.

3.2 Cockle fishing

Cockle fishermen operate in the Wadden Sea, Eastern and Western Scheldt and the Voordelta. Since 1974 all those using mechanical gears need a licence; initially 36 were issued increasing to 37 in 1992. Licences are saleable and in 1993 their market value was estimated at around Dfl 1 million. Of the total 33 are owned by a handful of large companies and the remaining four by skipper-owners. Investments in the cockle fisheries are considerable. Nearly all cockle landings are exported, principally to Spain and Portugal, after being cooked on board or in the shore based canneries. Though fluctuating significantly, the Wadden Sea stocks average around 50 million kg with an annual production of 6 million kg of cockle meat deemed acceptable. Employment in the sector including canning and trade - amounts to some 300 persons. Skippers and crew are paid either a flat wage or on a commission basis.

Out of season the cocklemen will meticulously map the locations with particularly abundant stocks, returning to exploit these areas during the fishing season. Like the musselmen, cockle fishers experienced particularly poor seasons in 1990 and 1991 when, due to natural causes including low winter temperatures and heavy storms, stocks were very low yielding an estimated 4 million kg of cockle meat. Despite this fishing continued resulting in serious stock depletion, censure from both scientists and environmentalists and the closure of a large area of the Wadden Sea and Eastern Scheldt.

In their fishing plans, the cockle fishermen agreed to reduce their harvesting capacity by reducing the number of hydraulic suction dredges from two per licence to one. A cockleman holding two licences could scrap one of his vessels and retain two gears on the other. As a result the number of vessels in the cockle fleet was reduced from 37 to 23. Use of hydraulic dredges is also subject to restrictions, while sorting machines on board can ensure that undersized cockles are returned live to the sea. During

September, October and November, the active fishing season, cocklemen are permitted to fish four days a week, deciding among themselves where each vessel will operate in an attempt to disperse fishing effort for the benefit of the fishermen and the stocks. The dedicated cockle PO, formed in 1995 and which incorporates all cocklemen using mechanical gears, will censure anyone not complying with the fishing plan. Some who breached the rules in the early days were fined and one skipper was dismissed by his company for persistent non-compliance (Steins, 1997). Today compliance rates are high.

So far, co-management seems to satisfy all the parties involved except the conservationists. But whereas the musselmen fared reasonably well in the 1990s, cockle fishers have faced huge problems. In 1996 and 1997, probably on account of the harsh winter of 1995-96, the cockle stock remained below the threshold level of an estimated 12.6 million kg of cockle meat: suction dredging has been banned in 26% of the intertidal areas of the Wadden Sea and the cockle fishery in the Eastern Scheldt completely shut down. Since 1993, landings have fallen below 7 million kg, reaching as low as 1.4 million kg in 1997. Subsequently, catches have increased considerably and were close to 10 million kg of cockle meat in 1998 and 8 million kg in 1999. Some cockle fishermen have begun catching *Spisula subtrunca* within the 12 mile zone and in 1997 sixteen firms established a PO drawing up a fishing plan similar to that for cockle fishing. The fishery is closed from December to May but otherwise unregulated. Returns are, however, much lower than in the cockle industry.

In addition to mechanical cockle fishing, some fishers rely entirely on handraking which until 1989 was unrestricted. But since the number of licence applications rose to over 350 a drastic reduction in numbers became necessary to prevent disturbance of the intertidal areas. Non-transferable permits were issued to 84 professional, small scale cocklemen on the basis that they were earning at least half the minimum wage through cockling. Today there are 77 licence holders of whom 25 are active and 20 depend on cockling for their livelihood. Most full time and some part time cocklemen have organised themselves in an association which acts as an intermediary between their interests and those of the traders, state, conservationists and the mechanical harvesting sector. As of 2000, five non-mechanised cockle fishers have become members of the PO. One eighteenth of the TAC is allocated to the hand operators; when the TAC is low, as in most recent years, the full time handworkers are forced to find alternative sources of income. There is a strict segregation of the cockle grounds with the handworkers allowed to fish year round. Since 1992 the sector has operated under a fishing plan, drawn up by their own association, signed up to by 30 fishers, but lacking in means of enforcement. Those who have opted out of the plan are only permitted to fish in one designated area. Unlike the larger, mechanised cocklemen, the handrakers have so far escaped the wrath of the conservation lobby.

3.3 Oyster production

For very many years oyster farming in Zeeland's coastal waters had been a lucrative business yielding between ten and thirty million marketable oysters each year. However, like cockles, oysters are vulnerable to low temperatures and following the drastically cold winter of 1962-3 a large proportion of the oyster stocks were killed. This, together with the threatened damming of the Eastern Scheldt, persuaded most to give up oyster farming. But the proposal to dam the Scheldt was abandoned in 1973 in favour of a storm surge barrier, leaving the nine surviving firms in a highly advantageous situation since they now leased nearly all the oyster beds. However, against the advice of the fisheries biologists, these firms decided to restock the beds with imported oysters from France which turned out to be infected with the parasitic disease, *Bonamia ostreae.* By 1980 the disease had affected stocks in the Eastern Scheldt and the state intervened to ban the cultivation of the European flat oyster.

Only in the Grevelingen, which had been dammed earlier but conserved as a saltwater lake, was the cultivation of *Ostrea edulis* continued. In the second half of the 1980s an annual yield of *circa* ten million oysters was harvested in the Grevelingen. New claimants demanded entry to the industry and after a political battle won their case. Since 1990 the state has rented plots to a total of 27 firms with rents, based on the estimated production value of the plot, yielding Dfl 1.3 million in state revenue in 1993. No new leaseholders are now admitted. Following the new fisheries policy in 1993, oyster plots have been redistributed. Earlier, in 1989, *Bonamia ostreae* was found to have spread to the Grevelingen: harvests declined sharply from around ten million to half a million oysters and this desperate situation continues to the present day although mortality from *Bonamia* appears to be on the decrease. Most of the oysters are exported to Belgium.

A profitable business can no longer operate on the basis of flat oysters alone. Another species - the cup or Pacific oyster (*Crassostrea gigas*) introduced in the 1970s - has proliferated in Zeeland waters and is now exploited by the existing oyster firms and 34 permit holders for the oyster fishery in the common area. Whereas flat oysters mature in six years, cup oysters are marketable in two years. In recent years, production has expanded from *circa* 10 million to 25 million oysters, even reaching a peak of 29 million in 1999. It is partly a capture and partly a culture fishery, but only the tenants of the cultivation plots are allowed to fish for oysters. The value of cup oysters is considerably lower than that of the European flat oyster, but since they are more prolific they are also serious competitors for food and space. Most oystermen combine oyster farming and fishing with other fishing or shellfish farming activities. In the first half of the 1990s, the value of oyster landings averaged Dfl 14.8 million a year, but by 1998 it had dropped to Dfl 6 million. A dedicated PO was established the following year and has 28 oyster producer members; together they account for 90% of Dutch oyster output.

3.4 Shrimp fishing

Some 233 Dutch vessels are licensed to shrimp within the 200 mile EEZ; their number is declining. Shrimping is done mainly in the Wadden Sea where according to a policy statement issued in 1988 the number of licensed shrimpers is restricted to 98 at least until the year 2000. Today 90 licences are extant: of these 46 are for boats permitted to fish within the 3 mile limits and 30 within the 12 mile zone. Wadden Sea shrimping is a year round activity though fishing is permitted on only four days of the week and may not be undertaken at weekends. A further 18 fishermen with historic beam trawling rights are allowed to catch shrimps in the Eastern Scheldt but only for part of the year. To protect the juvenile flatfish stocks, the shrimpers must meet stringent bycatch requirements, using sorting grids to separate the juvenile fish from the target catch of shrimps. Wadden Sea shrimp licences are transferable, changing hands for around Dfl 150,000; those in the Eastern Scheldt are not.

Unlike the mussel and cockle industries described earlier, shrimping has been scarcely affected by the change in fisheries policy in 1993. No new restrictive measures have been introduced. Indeed, shrimping is generally held to have only a relatively low impact on the natural environment. In the early 1990s landings averaged around 8 million kg rising to over 10 million in the middle years of the decade when the total value exceeded Dfl 60 million. In 1999 11 million kg was landed with a value of almost Dfl 85 million. Although shrimp processing is an important activity in the Netherlands, with Dutch firms handling around 60% of all European landings, most of the market for shrimps lies abroad.

Wadden Sea shrimpers are organised in the *Nederlandse Bond van Garnalen Kustvissers* (Dutch Union of Coastal Shrimpers); other shrimpers are members of the *Vissersbond* (Fishermen's Union). Most also join local or regional POs. The advantages of PO membership include not only access to minimum landing price schemes but also entry to the Biesheuvel self-management groups in the case of vessels with quota allocations for sole, plaice and roundfish. Individual shrimpers are not required to sign an agreement with the PO (as is the case with mussel and cockle fishermen) but are still subject to the discipline of the PO and can be fined for infringement of PO rules. A problem of freeriding occurs where non-members derive the benefits of a well regulated market without being subject to the PO's rules of engagement. As with the mussel fishermen noted above, there is a strong feeling among PO members of the need to extend PO rules to cover non-members.

One of the problems facing the full time shrimper is the tendency for the much larger cutter vessels to switch to shrimping when their demersal catch quotas are low. Quayside prices tumble when landings are significantly increased. Moreover, the cutters, not exceeding 300 hp, permitted to fish both within and outside the 12 mile zone and with quota entitlements to flatfish and roundfish species, have the advantage of greater flexibility over the shrimp boats restricted to inshore waters as a result of safety regulations and thus dependent on a single species. Expansion of the cutter fleet's harvesting capacity does cause concern over the possible overexploitation of shrimp stocks in the inshore zone.

3.5 Other fisheries

One measure of the supreme importance of the four key inshore fisheries - mussels, cockles, oysters and shrimps - is the almost complete lack of attention paid to other fisheries occurring within the inshore zone in the various policy documents and the paucity of statistical data. As in other European countries the official statistics do not distinguish between offshore and inshore harvests. Nonetheless a considerable variety of species - including eel, sprats, grey mullet, bass, lobster and anchovy - are caught using a range of gears, including fyke and set nets, pots and weirs, *inter alia,* not to mention the digging of lugworm in the intertidal areas of the Wadden Sea. However, land reclamation and the construction of dykes and dams in the Wadden Sea area in the 1930s and in the Zeeland area from the 1960s onwards have had devastating consequences for such species as herring, anchovy, turbot, ray, brill, periwinkle and whelk which once featured quite prominently in the mixed fisheries of the small scale inshore fleets in the Netherlands.

4. INSHORE FISHERIES MANAGEMENT: AN EVALUATION

The characteristic diversity of the inshore fishing industry makes for difficulties in creating an overarching policy that meets with the agreement and compliance of all inshore fishermen, let alone non-fishing interests. The tendency in the past had been to support those sectors which contribute most to the economy. Thus for a long time the mussel industry was favoured over shrimping, while mechanised cockle fishing received more benefits than the handrakers. Since the late 1980s, however, the order of priority has changed: nature conservation interests are now given first preference. In response to trenchant criticism of the shellfish industry and the early trilateral agreements concerning the management of the Wadden Sea, the state systematically set out to reconcile natural values and economic activities in its approach to fisheries policy in the period from 1993-2003. Co-management regimes were proposed for the inshore and offshore sectors. Mussel and cockle fishermen were encouraged to establish their own POs and to draft fishing plans in order to arrive at a state of self-management. Only the shrimping sector stood exempt from this new approach.

Positive effects of fishing plans in the mussel and cockle industries include improvements in the quality of shellfish landed and structural rationalisation through the concentration of licences. Shrimpers also enjoy the economic benefits of a licensing system since new competitors cannot enter the business. Black boxes used by cockle and mussel fishermen enable close monitoring of their fishing activities; when fishermen are accused of illegal fishing computer registered information is used to identify the offender. Moreover, the decision to disperse the cockle fleet avoids over-concentration of fishing in particular locations. In the mussel sector, mussel seed is used more efficiently; with the introduction of seed mussel quotas, fewer mussels are planted on the plots yet efficiency has increased threefold - less seed has resulted in outputs not dissimilar to the situation prior to the introduction of the quota system (Steins, 1997).

There are negative effects, however, including financial losses from decreasing output overall and more seriously from closures in the cockle sector. Cockle fishermen receive no compensation for their economic losses and an increasing ambivalence about the new policy threatens to undermine its legitimacy. For both cockle and mussel fishers the traditional freedom to deploy their experience and skill in a competitive form of fishing has been replaced by a system of PO controlled rights and obligations. Significantly, the cockle fishermen now undertake the annual mapping of resource rich locations collectively rather than individually.

The licensing system privileges those presently working in the industry and obstructs the entry of newcomers. Thus the firms tend to remain within a small network of kin: without the aid of family members who are vessel owners and licence holders, it is virtually impossible to become an owner operator. One major problem for the mussel industry is that not all musselmen are members of the PO and therefore do not have to comply with the fishing plans - a situation which can undermine the co-management regime where non-members can rely on alternative provisions made by the state. So far the Ministry has refused to grant an *extension de régime*. The same applies to the shrimping industry. Other than the exercise of informal social controls, no sanctions can be taken against those who undermine the formal agreements. Effective management would seem to require the government either to make PO membership obligatory, though this may contravene the articles of association, or to make the fishing plans binding on non-members.

The co-management regime operates within a framework set by government. It has increased the legitimacy of state-led policy, enabling an integrated approach to the management of coastal waters. Through their systems of self-management, cockle and mussel fishermen now perceive the government's approach as necessary and legitimate. Compliance with rules and regulations is high. They themselves refer to the period prior to the new regime as an irresponsible 'wild west' fishery. As long as the economic viability of their enterprises is not endangered, they are likely to accept the rules. But there is a potential threat: should they experience a prolonged period of financial recession, compliance will probably decline and pressures on the state institutions to relax the restrictive measures will increase.

There is another side to the success story. Many small scale entrepreneurs have been ousted from the industry while newcomers have failed to gain entry. There is, therefore, a large measure of social immobility, with most firms reliant upon occupational inheritance. The inshore industry now resembles a closed shop: outside the family networks, there are scarcely any new owner operators. Moreover, inheritance taxation creates difficulties for the survival of the family firm. The market value of quotas and transferable licences has become so high that only the wealthiest firms can afford to buy them.

5. EMERGING ISSUES: THE TENSIONS OF MULTIPLE USE

The key issues affecting inshore fisheries in the Netherlands are predominantly multiple use conflicts. First, in recent years conservation organisations have become stakeholders in the decision making concerning the inshore domain. They favour the non-utilisation of renewable marine resources and seek to turn areas like the Wadden Sea and the Eastern Scheldt into nature parks where no human activities are allowed. As claimants to the marine domain they enjoy widespread popular support and have achieved considerable political clout. Secondly, multiple use of the inshore domain for fishing leads to conflict between groups of fishermen whose activities cannot be carried out simultaneously. Where exploiting similar niches fishing methods based on stationary gears (fyke nets, set nets, fish traps) cannot cocxist with mobile gears (beam trawls, pair trawls, dredge nets). Similarly, mussel farming is carried out on private allotments in locations also preferred by cockle and shrimp fishermen and their modes of exploitation do not tally. Thirdly, inshore waters are increasingly used for purposes other than fishing and nature conservation. Holiday making, recreation, military activities, maritime traffic, dredging, oil and gas development, industrial and harbour construction and modern agriculture all interfere in some measure with inshore fishing and shellfish farming. The first two issues are considered in more detail below.

5.1 Fishermen v. environmentalists

Inshore fishermen have been subject to increasing interrogation by environmentalists regarding the impacts of their activities on marine habitats and ecosystems especially in the Wadden Sea where the scarcity of cockles and mussels in the early 1990s brought vehement criticism of the shellfish industry from the conservation lobby. The natural shellfish beds are seen as precious biotypes, but to satisfy the need for mussel seed and cockles fishermen must exploit them. The scarcity coincided with high mortalities among seabirds, especially eiders and oystercatchers: the environmentalists accused cockle and mussel fishermen of plundering the scarce supplies leaving nothing for the seabirds. Organisations including the Wadden Sea Association, Bird Protection and the World Wildlife Fund (WWF) sought legal action to prevent the state handing out permits to mussel seed and cockle fishers, but were unsuccessful. The sixth Trilateral Government Conference on the Protection of the Wadden Sea in Esbjerg 1991 decided on the permanent closure of large parts of the Wadden Sea for cockle and mussel fisheries. The Dutch fisheries minister had already decided to ban cockle fishing in the Wadden Sea for a prolonged period.

Subsequently the debate became mired in accusation and counter-accusation. Cocklemen, who had voluntarily agreed a reduction in fishing effort in 1992, in particular were vilified. According to the industry, fishermen always receive the 'disbenefit of doubt' (*NRC Handelsblad*, September 3, 1992), while the conservation lobby complained of the role of 'might as right' in respect of the fishermen's behaviour - self-imposed restrictions were seen as mere window dressing. The highly emotive confrontation tended to obscure the scientific explanation for the natural decline in mussel stocks.

Claims for the protection of marine wildlife have been recognised. Since 1993, fishermen's organisations and state representatives, in consultation with environmental organisations, have agreed to restrict fishing in the intertidal zone. In addition to the voluntary closures of 5% of the cockle grounds and 10% of the mussel grounds, 26% of the intertidal zone in the Wadden Sea and 14% in the Eastern Scheldt are closed to shellfishing, while the fisheries sector and government agencies cooperate in the management of fishing elsewhere in order to guarantee a sufficiency of food supply for the seabird population whatever the state of the stocks. Although the environmental organisations agreed the new management scheme, the underlying tension still simmers: the environmentalists continue to argue for the closure of all intertidal areas to the fishing industry. According to Bird Protection

> 'We should introduce a moratorium and agree to abandon fishing for mussels and cockles for a period of ten years. We must reserve the beds for nature.'
>
> (*de Volkskrant*, September 20, 1997)

For WWF the antidote to 'over management' of the Wadden Sea is to allow nature to take its course.

In contrast to the attempts to achieve consensus, some conservationists insist on steering a confrontational course in their attempt to frustrate the use of the inshore domain by shellfish fishermen and farmers. They can claim the tacit support of scientists at the Institute of Marine Research. Blame for the destruction of the natural mussel and cockle beds in the late 1980s and early 1990s is laid at the door of the fishermen and their deployment of suction dredges - significantly, the research was funded by the conservation organisations. The continued antagonism of conservationists is met with resentment by both fishermen's organisations and government officials.

> 'Though they claim not to oppose fishermen but policy, the effect is the same. Conservationists should well realise that they can achieve more with a policy for which there is support than with a policy that goes against the grain of fishermen.'
>
> (Bert Keus, in *Platform* 1997, 13 (11): p. 8-9)

What is clear from the controversy is the intensifying distrust which has to be resolved before any truly integrated management approach can be implemented. The eighth Trilateral Conference in 1997 reaffirmed the need for continuing efforts to conserve and protect the Wadden Sea and that the implementation of policies must be undertaken through cooperative dialogue with all stakeholders. Though fishermen's organisations and environmentalists may sit around the negotiating table, the latter appear unwilling to relax their aspirations. They have rejected the positive evaluation of the current policy undertaken in 1998 and sought legal action in the European Court to reverse the policy, but to no avail. For their part, mussel and cockle fishermen are also disappointed with the policy evaluation, pointing out that they are only allowed to fish in 3.5% of the Wadden Sea - a restriction which has cost them millions of guilders in lost income.

5.2 Fishermen divided

Musselmen confront shrimp fishermen in the north over the privatisation of shellfish plots which renders multiple use impossible. Since the 1950s shrimping territory has been drastically reduced. Shrimpers resent the fact that Zeelanders can plant mussels in 'their' territory. In 1994, Texel and Wieringen shrimpers had to relinquish 470 ha so that the state could allocate new mussel plots to replace unproductive ones. The shrimpers were offered compensation: some accepted while others contested the situation in court and lost. Musselmen have also offered to buy out shrimpers or pay compensation from their own funds to help shrimpers convert their boats and obtain permits to fish beyond the 12 mile limits.

Cockle fishers have recently attempted to expand their own areas of operation by occupying unworked mussel plots. Approximately 1,000 ha have been officially reallocated as a reciprocal for the allocation of new mussel plots. In their turn handrakers have struggled against the mechanised sector in terms of access to cockle grounds finally convincing the state authorities to reserve certain areas for their exclusive use. Meanwhile there is tension between Zeeland musselmen and oyster farmers over apparently unused oyster beds lying fallow in anticipation of the eradication of *Bonamia*. Quarrels have also occurred between oystermen who survived the collapse of stocks in the 1960s and those who decided to quit in the expectation that the Eastern Scheldt would be closed but felt cheated when the government later reversed its plans. Oystermen who had originally worked the Grevelingen but opted for mussel plots in the Wadden Sea or eel fishing rights in Grevelingen by way of compensation were aggrieved to see former competitors harvesting oysters in what they considered to be 'their' fishing grounds. After several legal actions, judgement was finally decided in 1987 in favour of allocating plots in the Grevelingen to 15 oyster firms - since when no new tenants have been admitted. One of the certainties of privatised shellfish farming is that established firms will invariably oppose the entry of newcomers and may resort to litigation to defend their rights.

6. CONCLUSIONS

Modern shellfish production, such as that dominating the inshore fisheries in the Netherlands today, implies an increasing level of intervention by man, which in turn leaves little room for multiple use. Success in shellfish farming is usually achieved at the cost of other fisheries, through the exclusive occupation of marine space formerly shared with other fisheries. Moreover, privatisation implies the selfish protection of individual interests, the rejection of claims by potential new entrants and can lead to the marginalisation of small scale fishers. These internal conflicts and problems of equity need to be solved or friction will increase. External pressures resulting mainly from permanent or seasonal urbanisation - until now effectively restrained by vigorous physical planning - cannot be held back permanently. The moratorium on oil and gas exploration in the Wadden Sea may be lifted in the future and it will be surprising if the shellfish industry does not have to cede some of its territory to tourism development at some point in the near future.

Apart from internal problems of social equity, monoculture can also lead to ecological issues - a point forcefully recognised by the environmental lobby. Considerable fluctuations in natural stocks can occur without any apparent relationship to stock recruitment. The dilemma is that shellfish cultivation can enhance production, but only at the cost of ecological diversity. Carelessness on the part of a few users or forces beyond man's control can seriously deplete the common resource and erode the collective interests of all stakeholders. In recognition of these dangers various partners in the Netherlands have come together to create a unique system of co-management intended for the mutual benefit of fishing and wildlife conservation. The legitimacy of this system is, however, at risk of being undermined either through persistent denigration of its achievements by certain environmental groups or as a result of persistently poor economic returns caused by what some may consider to be an overregulated inshore fishing industry.

7. REFERENCES

Beheersplan Schelpdiervisserij (2000) *Beheersplan Schelpdiervisserij Kustwateren 1999-2003*, Productschap Vis, Rijswijk.

Keus, B.J. (1997) Co-management in Dutch Shellfish Fisheries, *Wadden Sea Newsletter*, **2**, 9-12.

Planologische Kernbeslissing (1981) *Planologische Kernbeslissing Waddenzee,* SDU, Uitgeverij, The Hague.

Steins, N.A. (1997) *Balancing fisheries and nature: three case studies of fisheries co-management in the Dutch Wadden Sea*, Working Papers in Coastal Zone Management 24, Centre of Coastal Zone Management, University of Portsmouth, Portsmouth.

Steins, N.A. (2000) Co-managing fisheries and nature in the Dutch Wadden Sea, in L. Heaps (ed.), *Integrating Biodioversity and EU Fisheries Policy: Rebuilding Healthy and Productive Ecosystems Vol 2 Workshop Reports*, WWF, London, pp. 41-44.

Structuurnota (1998) *Structuurnota Zee- en kustvisserij. Evaluatie van de maatregelen in de kustvisserij gedurende de eerste fase (1993-1997)*, Ministerie van Landbouw, Natuurbeheer en Visserij, The Hague.

Vissen (1993) *Vissen naar evenwicht. Structuurnota Zee- en kustvisserij*, Ministerie van Landbouw, Natuurbeheer en Visserij, Den Haag.

Visserij in cijfers (1999) LEI, Afdeling Visserij, The Hague.

Chapter 6

GREAT BRITAIN

Jeremy Phillipson and David Symes

1. INTRODUCTION

Given the long and elaborate coastline of the United Kingdom (Figure 6.1) - and with a baseline in Scotland which encloses the extensive area of the Minch - it would be surprising if inshore fisheries were not to play an important role in the UK's fishing industry, as indeed is the case. However, as with many European countries, it is no longer easy to define with any precision what is meant by the term inshore fisheries. We know, for example, that *circa* 80% of the registered fleet comprises vessels under 12 m in length and we can infer that not far short of half the active fishermen are involved within the small boat sector with perhaps a third of them engaged on a part time or seasonal basis. But we also know that, as a consequence of technological developments, much of this so-called inshore fleet is capable of operating well beyond the 6 mile limits which have traditionally marked the boundary between inshore and offshore waters. Moreover, these inshore waters are also exploited by non-local, nomadic vessels much larger in size than the under 12 m inshore boat.

Even if vessel size or territorial limits were able to provide an unambiguous definition of inshore fisheries, official statistics would offer little help in assessing their overall economic contribution. Separate landing data are not available for the under 12 m fleet nor for catches taken within the 6 (or 12) mile zone. Thus, any attempt to measure the economic significance of the inshore sector has to rely on informed guesswork.

In Britain, inshore fisheries are based largely, but by no means exclusively around the shellfisheries. As Figure 6.2 indicates, these have proved a strong growth sector in UK fisheries, increasing by 35% in terms of volume and 66% in value between 1989 and 1998, thus helping to offset declining levels of activity in the demersal and pelagic fisheries. By 1998 the

D. Symes and J. Phillipson (eds.), Inshore Fisheries Management, 97–118.

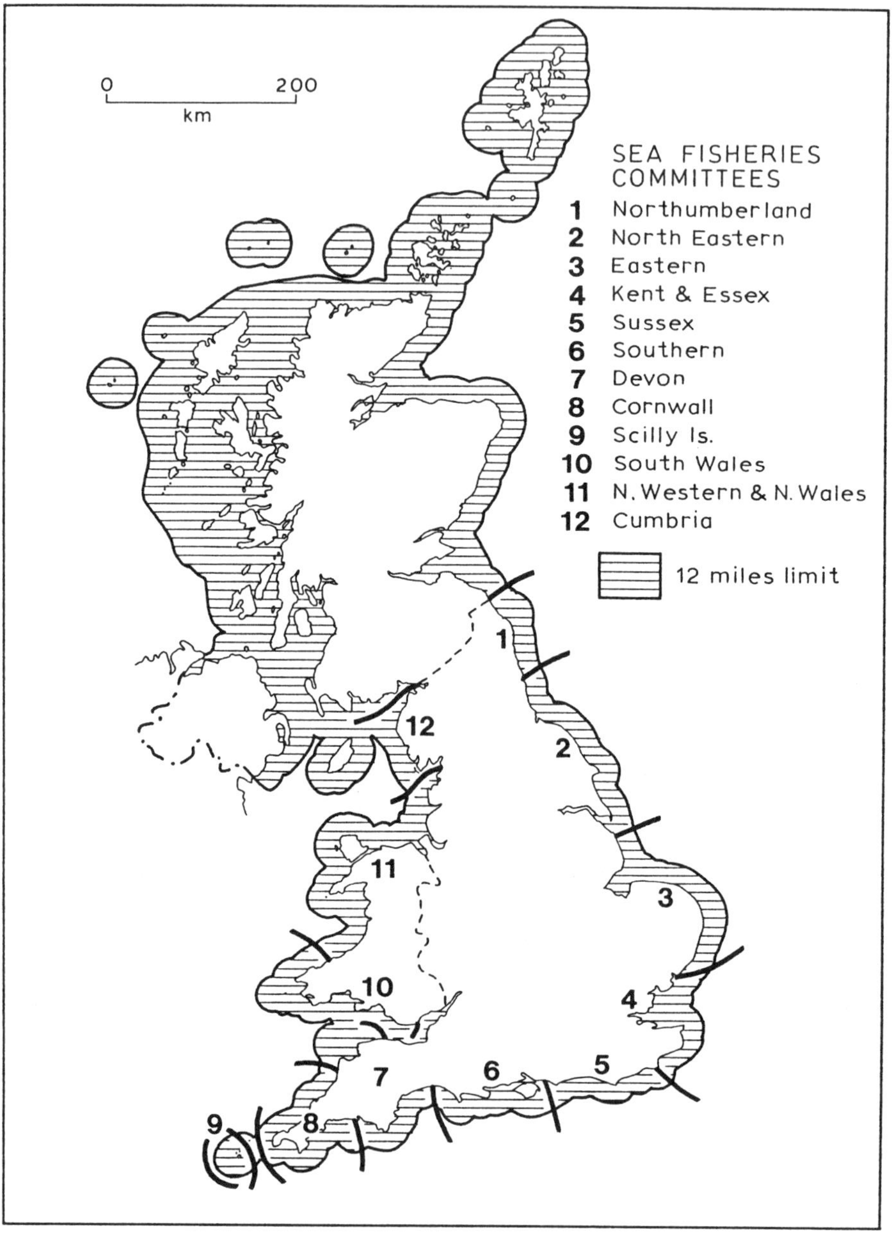

Figure 6.1 UK Inshore Waters, Showing Boundaries to Sea Fisheries Districts in England and Wales (0-6 nm Only)

shellfish sector accounted for a third of the value of landings into UK ports. By no means all shellfish landings can be attributed to the inshore sector.

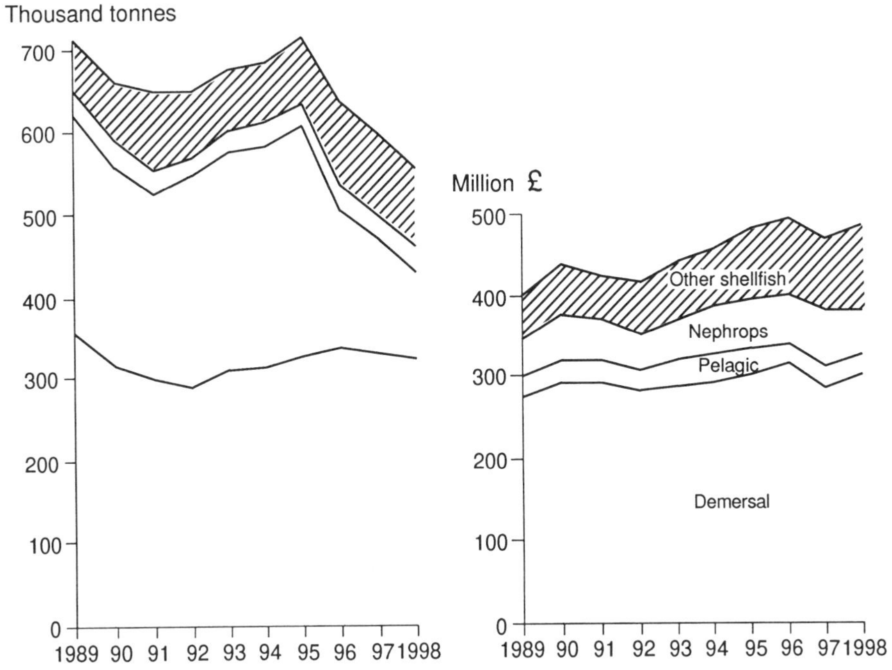

Figure 6.2 UK Landings, 1989-1998, by Volume and Value (Source: Sea Fisheries Statistics, Various Dates)

Nephrops, a particularly strong growth area throughout the 1990s, is essentially an offshore trawl fishery. Other traditional shellfish species are now being taken well outside the 6 mile limits. Allowing for these non-inshore activities - and for the comparatively small allocations of quota regulated demersal and pelagic species reserved for the under 10 m fleet - it would seem reasonable to suggest that inshore fisheries account for somewhere between a fifth and a quarter of the total value of UK landings.

As will become evident from Figures 6.3 and 6.4, the relative importance of the shellfish sector - and thus presumably of the inshore fisheries - varies quite markedly between England and Wales on the one hand and Scotland on the other. It assumes a much greater significance south of the border; not only is the influence of *nephrops* quite negligible, but the shellfish sector accounts for appreciably more in absolute and relative terms. In Northern Ireland, the comparatively small fisheries sector is surprisingly dominated by a single fishery, with *nephrops* accounting for 46% of the value of all landings.

There is a second more plausible reason for paying relatively more attention to the management of inshore fisheries south of the border.

Although, in general, fisheries management throughout the UK has been subject to parallel legislation with nominally separate Acts and statutory instruments (Orders) regulating the industry both north and south of the border, in the matter of inshore fisheries management there has been a long history of separate development. While in England and Wales a unique system of devolved management for inshore fisheries has been in place for well over a century, in Scotland - and again in Northern Ireland - management responsibility has been located centrally in the Scottish (and Northern Ireland) offices. Since devolution in 1998, and confirmation of the Scottish Parliament and Scottish Executive's overall responsibility for fisheries up to 6 nm from the baseline, there has been some movement in Scotland to increase the level of co-management. Notwithstanding this, the systems of management north and south of the border remain highly distinctive.

2. INSHORE FISHERIES IN ENGLAND AND WALES

Since the collapse of the distant water fisheries in the late 1970s following the establishment of Exclusive Economic Zones (EEZs) throughout the North Atlantic, the polarisation of the fishing industry in England and Wales between the capitalistically structured, company owned trawler fleets of the larger ports of Hull, Grimsby, North Shields, Fleetwood and Milford Haven and the family based, artisanal 'day' boats dispersed among several scores of small fish harbours, has evaporated. Today the bulk of the fleet is designed to operate in near and middle waters and, according to a 20% sample survey conducted in 1996, 56% of all active fishermen in England and Wales are to be found in the under 10 m sector.

In general, the significance of the inshore fisheries - whether measured in terms of vessel size or the relative importance of shellfish landings - increases as one moves southwards and away from the larger ports historically associated with the deep sea trawling industry. The stronger concentrations of inshore boats are found along the south coast from the Thames to the Severn estuaries.

Figure 6.3 indicates the stabilising influence of the shellfish sector at a time of gradual decline in demersal landings and strong fluctuations in the pelagic fisheries. By 1998 shellfish, excluding *nephrops*, valued at £60.3 million, accounted for 35% of the value of all landings in England and Wales. The most important and widespread shellfish species are crabs (£20.4 million) - placing it second behind cod in the rank order of landings by value - scallops (£13.9 million) and lobsters (£8.1 million), together with locally

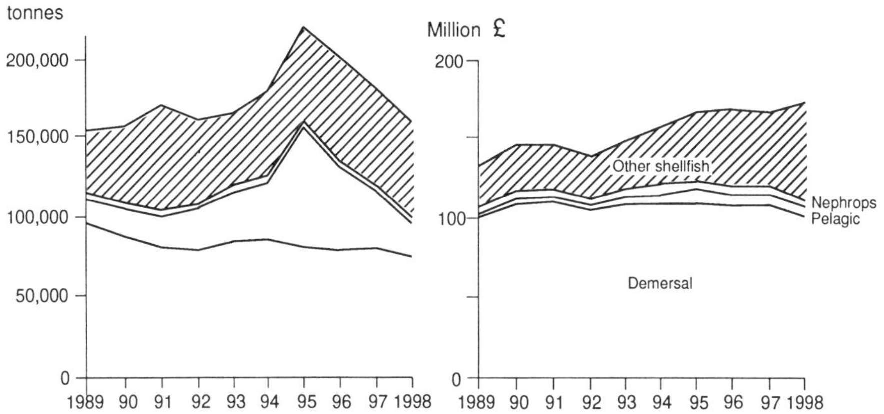

Figure 6.3 Landings in England and Wales, 1989-1998, by Volume and Value (Source: Sea Fisheries Statistics, Various Dates)

significant concentrations of bivalves (cockles, mussels and oysters). Apart from shellfish, the inshore fisheries include participation in demersal fisheries (especially cod), a strong regional interest in handlining for mackerel in the south west, and local interest in salmon netting along the north east coast. The latter is gradually being phased out through a progressive reduction in the number of licences, ostensibly to protect the considerable financial interests of the inland sports fishery for salmon in Scotland.

The broad spectrum of inshore species is matched by an equally wide range of fishing gears (Table 6.1). Generally inshore fisheries remain characterised by varying combinations of several clearly defined fishing seasons with each one frequently involving only a few weeks activity. On the north east coast, for example, a typical combination for an under 10 m boat fishing for 45 to 48 weeks of the year might involve crab or lobster potting throughout the period from March to late October, overlapping with a twelve week salmon netting season (May-August), and a weather dependent autumn and winter longlining or gillnet fishery for cod.

Each unit is, therefore, a multi-specialist enterprise reliant upon highly developed local knowledge and a diverse range of professional skills. Recently, however, there has been a tendency among a small but growing number of boats to specialise in a single fishery on a year round basis. Specialisation is symptomatic of an increasing culture of professionalism within the inshore fisheries especially among the younger fishermen. The sector is certainly not immune from technological change nor from the tendency towards increasing intensification of fishing effort. Along the Yorkshire coast, for example, many lobster boats have expanded their

Table 6.1 Inshore fisheries and gear types in England and Wales

Species group	Species	Gear
Demersal	cod, whiting, haddock, saithe, pollack, dogfish	demersal otter trawl, pair trawl, seine net, gillnet, trammel net, longline, handline
	Dover sole, plaice, flounder, turbot, monkfish, rays	beam trawl, demersal otter trawl, seine net, tangle net, trammel net, longline
Pelagic	mackerel, herring, sprat, bass, mullet	gillnet, pelagic trawl, handline, beach seine, trolling
Migratory	salmon, sea trout, eel	gillnet, beach seine, trap, hand held nets, otter trawl, fyke nets
Shellfish	lobsters, crabs, crawfish, nephrops, shrimp	pot, tangle net, beam trawl, demersal otter trawl
	cockles, mussels, scallops, oysters, clams, whelks, periwinkles, cuttlefish, squid	dredge, trawl, pot, hand gathering

Based on Gray, 1995

potting activities well beyond the 6 mile limits and extended the potting season to what is virtually an all year round activity using the more efficient parlour pots, with the result that landings have increased very significantly over the past few years.

3. INSHORE FISHERIES MANAGEMENT

The legislative basis for inshore fisheries management in England and Wales is embodied in three principal Acts dating from the 1960s (see Table 6.2), together with subsequent amendments providing for the incorporation of marine wildlife conservation concerns within fisheries management in the 1990s and specific provisions for salmon fisheries.

In effect the legislation lays the foundations for an integrated two tier system of management involving both centralised and quasi-autonomous regionally devolved approaches. Nationally, the Ministry of Agriculture, Fisheries and Food (MAFF) is responsible for the implementation of EU policies, in particular deciding on the detailed system of management for quota regulated species, and for the adoption of additional measures in respect of fisheries within the 12 mile zone. Significantly, under 10 m

Table 6.2 Legal instruments with particular influence on inshore fisheries in England and Wales

Legislation	Relevance to inshore fisheries
Salmon and Freshwater Fisheries (Protection) Act 1951	Prohibits poaching of salmon within 1 mile of the low water mark.
Sea Fisheries (Regulation) Act 1966	Establishes Sea Fisheries Committees (SFCs) in England and Wales. Appointment by SFC of sea fisheries officers to enforce SFC byelaws. Consolidates 1888-1930 Sea Fisheries Regulation Acts.
Sea Fisheries (Shellfish) Act 1967	Minister may make, by order, byelaws for establishment or regulation of shellfish fisheries. Holders of orders have exclusive rights of deposition, propagating, dredging, fishing for and taking shellfish. Consolidates previous enactments.
Sea Fisheries (Conservation) Act 1967	Consolidates provisions for the regulation of sea fisheries and authorises measures for improvement of resources e.g.: restrictions on commercial use of undersized fish; regulations of nets and gear; licensing of fishing boats; powers to restrict fishing by order of the Minister; powers to take measures to increase or improve resources.
Salmon and Freshwater Fisheries Act 1975	Regulates obstruction to the passage of fish up rivers and defines offences for obstructing this passage etc.
Salmon Act 1986	Mostly applies to Scotland. Establishes Salmon Fishery Districts. Establishes salmon fishery boards.
Water Resources Act 1991	The National Rivers Authority (now the Environment Agency) has duty to maintain, improve and develop salmon, trout and eel fisheries within 6 nm from baselines.
Sea Fisheries (Wildlife) Conservation Act 1992	Requires Ministers and relevant bodies to have regard to the conservation of fauna and flora in discharging their functions.
Conservation (Natural Habitats, &c.) Regulations 1994 (SI No.2716)	Regulations for implementation of the Habitats Directive.
Environment Act 1995	New environmental powers for SFCs; establishment of Environment Agency

vessels are excluded from the main sectoral quota system administered by the producers' organisations (Phillipson, 1999). Instead, a separate quota allocation based on aggregate landing data and established fishing patterns in each of the ICES areas, is managed centrally by MAFF. Individual vessel quotas are not allocated and the under 10 m fleet thus fishes against consolidated quotas until they are exhausted and the fishery closed. Since 1993 all under 10 m vessels must have a general licence to fish which in theory permits considerable freedom of location, target species and catch quotas; but in practice the inshore fleet is subject to very much closer regulation through the system of devolved regional management.

Devolved responsibility is divided, very unequally and somewhat uneasily, between the 12 Sea Fisheries Committees (SFCs) and the Environment Agency (EA), with the latter concerned essentially with the migratory species - principally the anadromous salmon and the catadromous eel - but also acting *in loco* SFC in a handful of English and Welsh estuaries, notably the Severn, Dee and Taw-Torridge. The EA is advised in matters of fisheries policy by a series of Regional Fisheries Advisory Committees and these, in turn, are assisted by local committees comprising different fishing interests including riparian owners, anglers, commercial netsmen and fish farmers. It has powers to limit the number of licences for salmon and eel fisheries and to frame byelaws to regulate fishing activity.

Sea Fisheries Committees, first established in 1888, are now governed by the *Sea Fisheries (Regulation) Act 1966* which confirms the statutory role of the 12 SFCs in England and Wales with jurisdiction extending originally from the low water mark to the 3 mile limits but subsequently amended to 6 miles in 1992. Thus the SFCs have powers to regulate fishing throughout the whole of the inshore zone from which non-coastal state vessels are barred under the current EU derogation. Regulation of UK waters between the 6 and 12 mile limits, where foreign fishing vessels are allowed on the basis of historic fishing rights, remains the responsibility of MAFF.

The Committees' management powers derive from two very different types of instrument: byelaws and Regulating Orders. SFCs are empowered to establish byelaws, on confirmation of the Fisheries Minister, for the purpose of regulating fishing methods and restricting or prohibiting sea fishing. In effect they permit an important degree of fine tuning of European and national legislation and the imposition of additional restrictions. In line with the specific needs of the local fisheries, SFCs are able to create their own menus of detailed restrictions and specifications. These commonly relate to the principle of zoning by, for example, limiting the access of vessels using towed gears to waters more than 3 miles from the shore, according to vessel size, thus implying a preference for smaller vessels deploying static or passive gears. Other byelaws may relate to the rigging of gears, mesh sizes, minimum landing sizes, partial or complete restrictions on particular fishing methods and the introduction of closed areas to protect spawning and nursery populations. Byelaw powers do not, however, extend to the introduction of licensing nor to the imposition of output controls (quotas).

They therefore stop short of empowering SFCs to adopt the two most commonly deployed means of modern fisheries management.

It was the *Sea Fisheries (Shellfish) Act 1967* which provided SFCs with possibly their most powerful management tool, certainly in respect of the important molluscan and - since the amendment in 1997 - crustacean fisheries with the introduction of Several and Regulating Orders. Several Orders refer specifically to the more sedentary molluscan fisheries. They remove the public right to fish through the granting of exclusive husbandry rights for 'depositing, dredging, propagating and taking shellfish' to individuals, groups of fishermen and other bodies, including SFCs. The grantee is able 'to make and maintain shellfish beds, to collect shells and remove them from one place to another' and is permitted to sublease portions of the sea bed to individuals or groups to lay and cultivate shellfish. Such rights to work what are in practice marine allotments are granted for a maximum of ten years and limited to a total of ten hectares.

While Several Orders offer a route to a strictly limited 'privatisation' of the fishery and the opportunity for individual responsibility in maintaining and enhancing the resource base, Regulating Orders provide the basis for effective regulation of relatively extensive shellfish grounds. The two largest Orders embracing the Thames Estuary cockle fishery and the Wash mollusc fisheries cover 116,000 and 69,000 ha respectively. Usually, granted for up to 30 years, Regulating Orders enable public bodies like SFCs to introduce restrictive licensing and to develop comprehensive management schemes through the attachment to the licence of specific conditions relating to effort and output limitations. Moreover, Regulating Orders also authorise the levying of licence fees, with the stipulation that licensing revenue is reinvested in the fishery through research and stock enhancement schemes.

Importantly SFCs have both the authority and the capacity through their own patrol boats to enforce local regulations and to help police national and European legislation within the 6 mile limits. They must, however, rely on prosecutions in the civil courts to bring offenders to book.

Inshore fisheries management in England and Wales is based, therefore, on two important principles. In the first place, the mosaic of access agreements created through byelaw regulations, Regulating Orders and Several rights involves a degree of partitioning of inshore waters between different fisheries, gear groups and vessel categories. Its objective may be to avoid conflicts arising in a situation characterised by many different modes of fishing competing for the same restricted space. Its effect, on the other hand, is the emergence of a system of local preference as the result of byelaws which favour small boat fisheries, controls over the granting of heritable licences and, most transparently, through the granting of Several Orders (Symes and Phillipson, 1998). To some this is very much in line with the original intention of the access derogation; to others it smacks of protectionism.

Perhaps more importantly, the system of devolved management allows for a high degree of sensitivity to local conditions in the handling of inshore fisheries. As a result, a remarkable diversity of regulation has evolved. Take, for example, the lobster fishery subject to a minimal level of management intervention at the EU or national level but where at the level of the Sea Fisheries Districts a very complex permutation of measures now occurs (Table 6.3). It is a moot point whether such diversity has a really compelling basis in science. Table 6.3 also serves to emphasise the point that inshore fisheries in England and Wales must be among the most heavily regulated in Europe at least.

Because of their particular functions in respect of inshore fisheries, SFCs also find themselves consulted formally and informally on a wide range of issues affecting the coastal zone. In recent years they have assumed an increasing role in relation to marine environmental conservation partly as the result of new legislation reshaping their formal responsibilities. The *Sea Fisheries (Wildlife Conservation) Act 1992* required all fisheries regulators, including SFCs, to 'have regard to the conservation of marine flora and fauna' when exercising their statutory duties and developing new byelaws. But it still left SFCs unable to restrict fishing solely on the grounds of environmental protection. The *Environment Act 1995* redressed this omission by empowering SFCs (and the EA) to develop byelaws expressly for purposes of environmental management and enabling them to regulate fisheries in order to maintain or restore areas to favourable conservation status. One of the first examples of these new byelaws was that drafted by the Cornish SFC in 1999 for the protection of seabirds involving the suspension of gillnetting in St. Ives Bay when seabird mortality through entanglement exceeds a predetermined level over a consecutive five day period.

A more specific recognition of the changing role of SFCs came with the implementation of the EU's Habitats Directive (Council Directive 92/43/EEC) and the establishment of a network of marine Special Areas of Conservation (SACs) within inshore waters (see Chapter 14). To date some 40 SACs have been identified around the UK coast - of these, the majority are off the coasts of England and Wales. Not only are SFCs recognised as 'relevant and competent authorities' with a duty to ensure that fishing activity is compatible with any necessary conservation measure but the Minister can also direct such authorities to formulate management plans for the SACs. Many 'joint management groups' have been formed with SFCs included, though only rarely have SFCs offered to undertake the role of lead agency.

Table 6.3 Variations in the management of lobster fisheries in inshore waters of England and Wales

	EU	UK	Northumber-land	North Eastern	Eastern	Kent and Essex	Sussex	Southern	Devon	Cornwall	Isles of Scilly	South Wales	North Wales and North West	Cumbria
Minimum landing size (mm)	85	87							90	90	90	90	90	
Maximum landing size												✓		
Ban on landing of soft lobsters			✓		✓									
Ban on landing of berried females					✓	✓			✓	✓				
V-notching			✓	✓					✓			✓	✓	
Escape vents in pots										✓				
Permit scheme (landing records)			✓	✓	✓			✓		✓		✓	✓	✓

Source: based on Addison, 2000

Although all 12 SFCs are governed by the same basic rules laid down in the principal legislation and must therefore undertake broadly the same range of statutory responsibilities, they vary very considerably in terms of geographical extent, organisational complexity and financial strength. The structural complexity derives in part from the number of coastal local authorities involved. Several SFCs are simply structured, identified with a single county and cover only a relatively short coastline: in these cases the committee is likely to number around 20 persons. Exceptionally a SFC may embrace ten or more local authorities, cover a correspondingly longer coastline and comprise a very much larger membership.

In similar fashion the size of the establishment of individual SFCs will vary according to the length of the coastline and the scale and complexity of the inshore fisheries. Moreover, the imposition of new duties, especially in the area of marine conservation, has added to the pressure to expand and diversify the staffing of SFCs. In the case of the North Eastern SFC covering some 230 km of coastline from the Tyne to south of the Humber estuary, staffing comprises a Chief Fishery Officer, four shore based fisheries officers, an environmental officer and four members of the patrol boat crew.

Funding for the SFCs - to cover staffing, the purchase, maintenance and operation of the patrol boat and most administrative costs - comes primarily from the constituent local authorities according to a precept based in part on the length of coastline and the relative importance of the local inshore fishing industry. With continuously rising costs for both manpower and the operation of the patrol boat and as a consequence of local authority reorganisation in the 1990s, adequate funding to ensure effective management of the inshore fisheries has become an increasing problem. Nonetheless at a total cost for England and Wales of around £5 million per annum, the SFCs offer a remarkably inexpensive and cost effective option.

Half of the available seats on the committees are allocated to elected members of the constituent local authorities and shared more or less in line with the level of their financial commitment. Of the remainder, one seat is reserved for the Environment Agency and the rest are filled by persons appointed by the Minister acting on advice from local fishermen's associations and MAFF's own District Inspectors. The appointees are intended to reflect the nature of the local inshore fishery and may include active or retired fishermen, processors and merchants, sports fishing interests and scientists drawn from local universities. At least one of the appointees must represent the environmental conservation interest and is commonly a member of a local wildlife trust.

Despite the fact that England and Wales can boast an almost uniquely comprehensive system of inshore fisheries management, which rests very largely on the strengths of the SFCs, there is a growing sense of the need to review existing arrangements. SFCs have been in existence for well over a hundred years and their basic constitution determined by the 1966 Act, since when there have been considerable changes to the circumstances of inshore fisheries and to the legal responsibilities conferred on SFCs through subsequent legislation. There is, however, no intention behind the growing insistence on a review of inshore fisheries management in England and Wales that the underlying concept of devolved management is redundant, nor that the value of SFCs is in doubt. Rather the need is to strengthen the system and ensure it is adequately equipped to cope with the kind of demands likely to be imposed upon it during the early years of the twenty first century. This certainly is the tenor of the recommendations from the recent House of Commons Committee report (1999) and from the internal report produced by the Association of Sea Fisheries Committees (2000).

Five key issues are likely to dominate discussions over the next few years, especially in the anticipation - and eventual realisation - of the renewal of the existing derogation relating to inshore waters:

- The extension of the SFCs' (and, where relevant, the EA's) jurisdiction from the current 6 to a possible 12 mile limit, implying either the removal of existing access agreements in the 6-12 mile zone and the extension of the ban on foreign fishing vessels throughout the consolidated zone or the granting of powers to the coastal state to impose regulations on both national and foreign vessels fishing within the 12 nm limits.

- Securing the long term future funding of SFCs; although very few are currently at risk in the short term, there is evidence that their work is being hampered by the limited level of funding available from the local authorities. Reform of funding, however, suggests a degree of direct funding from MAFF which is bound to raise questions over the future independence of SFCs.

- Restructuring the membership of SFCs; the current situation is bound up with the principle of democratic accountability at the local level and a recognition of the local authorities' role in funding the SFCs. The need to accommodate an increasingly wide range of stakeholder interests has diluted representation from the fishing industry itself - any financial restructuring could open up the opportunity to recalculate the proportional value of the different constituencies of membership.

- Creating new regulatory powers to allow SFCs to act proactively, rather than be forced into a catch-up game, in order to assert proper management control over both existing and new fisheries within inshore waters; the lack of precautionary power needs to be addressed and consideration also given to the inclusion of emergency powers to allow SFCs to respond quickly to unforeseen events.

- Enacting new primary legislation to consolidate the *ad hoc* adjustments to the SFCs' responsibilities since the 1966 Act and redefine the role and functions of SFCs in the light of these amendments. Previous requests for new primary legislation to enable SFCs to function efficiently and effectively have been rebutted on the grounds of insufficient parliamentary time. Reconfirmation of the inshore waters' access derogation, with or without amendment, could provide the catalyst for drafting appropriate new domestic legislation.

4. INSHORE FISHERIES AND THEIR MANAGEMENT IN SCOTLAND

While Scotland clearly dominates the overall pattern of landings in terms of volume and value, the emphasis on inshore fisheries is rather more subdued. As Figure 6.4 indicates, shellfish landings (excluding *nephrops*) are much lower at some 35,400 tonnes; valued at £42.3 million they account for only 15% of the gross earnings of the Scottish fleet, compared with 35% in the case of England and Wales. National statistics for Scotland, which reflect the pre-eminence of the larger demersal ports concentrated in the Grampian region of north east Scotland, tend to disguise the regional importance of inshore fisheries especially in the Highlands and Islands and the dispersal of inshore fishing activity among a large number of smaller ports around the Scottish coast. In general, the patterns of inshore fishing activity are broadly similar to those in England and Wales, with a combination of demersal fishing using both towed and static gears and potting and dredging for shellfish; the range of demersal species is generally narrower than in England and Wales.

Following parliamentary devolution, the application and enforcement of fisheries regulations within the Scottish zone of British fisheries limits, and therefore responsibility for inshore fisheries management, rests with the Scottish Executive Rural Affairs Department, formerly the Scottish Office Agriculture, Environment and Fisheries Department. In contrast to the well established system of Sea Fisheries Committees, under the *Inshore Fishing (Scotland) Act 1984*, responsibility for Scottish inshore fisheries

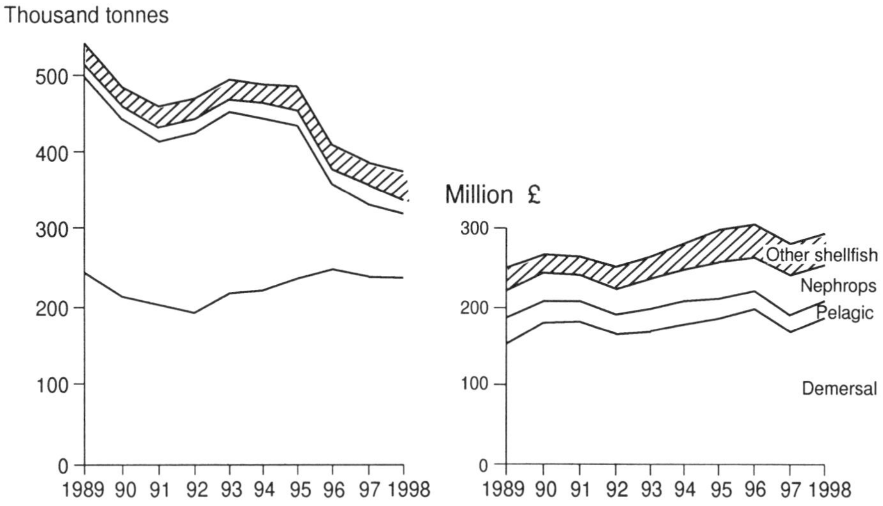

Figure 6.4 Landings in Scotland, 1989-1998, by Volume and Value (Source: Sea Fisheries Statistics, Various Dates)

management is held centrally by the Secretary of State, who has powers to prohibit sea fishing in specified areas by specified methods, or for particular descriptions of sea fish in Scottish waters. In practice, since the 1984 Act, there have been several Orders laying down specific management arrangements within the 6 nm limits. The principal Order is *The Inshore Fishing (Prohibition of Fishing and Fishing Methods) (Scotland) Order 1989* which has been amended on a triennial basis in consultation with the fishing industry.

For many years, and at least since the 1970s (Cameron Report, 1970), there has been a simmering debate over the potential for local management systems in regulating inshore fisheries in Scotland. Opposition to the development of statutory local management structures has been founded on a conflict of interest between the local small boat sector, represented primarily in the north west of Scotland and mobilised within the West of Four Fisheries Management Group and the Federation of Highlands and Islands Fishermen, who are in favour of some form of local management, and the more nomadic capital interests in the north east, which perceive moves towards local management as signifying their potential exclusion from traditional fishing grounds. Those in opposition are represented by the main power base of Scottish fisheries, the Scottish Fishermen's Federation and, coupled with a traditional preference on behalf of the Scottish Executive for voluntary rather than statutory local management arrangements (Scottish

Office, 1996), these have formed formidable opponents to local management.

The lack of local fisheries management structures in Scotland has been identified in several quarters as being detrimental to the effective management of inshore fisheries and the marine environment (Harrison, 1993; Seafish, 1991). Indeed, in the late 1990s there was a growing acknowledgement that some form of local management was appropriate for the inshore fisheries of Scotland and this led to the emergence of various local provisions. For some time, for example, Area Access Management Committees have been in place on the west coast of Scotland which serve to encourage the resolution of conflicts between mobile and static gears. More recently, the Scottish Executive's own *Programme for Government* included the explicit objective of encouraging greater local involvement in inshore fisheries (Scottish Executive, 2001). Such an initiative is to be progressed through two main approaches: firstly, via a new Scottish Inshore Fisheries Advisory Group (SIFAG), an advisory body set up in September 1999 to inform the development of inshore fisheries policy. SIFAG comprises representatives from fishermen's organisations, the Sea Fish Industry Authority, Scottish Natural Heritage, Highlands and Islands Enterprise and the Convention of Scottish Local Authorities. Its agenda has so far ranged widely and has included a review of management arrangements for 10 m and under vessels, consideration of shellfish licensing and a review of inshore fishing prohibitions.

Secondly, there is increasing emphasis being placed on the potential of local Several and Regulating Orders in Scottish fisheries. In May 2001 there were already 7 Several Orders in place in Scottish inshore waters with a further 10 applications for new Orders posited with the Scottish Executive. Regulating Orders, which arguably offer greater scope for local management, are more recent phenomena in Scotland. So far there is a single Regulating Order in place. This is based around Shetland and controlled by the Shetland Shellfish Management Organisation (SSMO). Following an original proposal in January 1998, the Order was granted in March 2000 for a nominated group of shellfish species within the 6 nm limits with the aim of maintaining and improving shellfish fisheries. SSMO, a non-profit making company, has a broad constituency including representatives from the local fishermen's organisation, regional council, community council, fish processing association, local fisheries college and the statutory conservation organisation in Scotland. Central to the management plan is a restrictive permit scheme, open to fishermen who are able to demonstrate a track record of fishing off Shetland.

Following the lead taken by Shetland, there is growing interest in the development of Regulating Orders elsewhere around the Scottish coast. An application for an Orkney Islands Order has already been submitted and there are other initiatives evolving in Fife, the Solway Firth and the Western Isles. Finally, there are proposals for a very extensive Order covering the inshore fisheries from Lochaber on the west coast of Scotland to Nairn on the east coast, subdivided into a series of separate sub-areas. All these initiatives, including the existing Shetland Regulating Order, represent important developments in Scottish inshore fisheries management. On the one hand, they raise questions concerning the changing nature of fisheries governance in Scotland following devolution. On the other hand, they pose several issues for administrators and local management interests alike. For example, it will be important to consider whether Several and Regulating Orders serve as inclusionary or exclusionary forces within local inshore fisheries and the extent to which they are able to mediate local interests. In addition, serious questions remain as to the ability of local organisations to enforce management measures in the absence of an independent enforcement capability at sea and the absence of stop and search powers similar to those held by the Sea Fisheries Committees south of the border.

5. AQUACULTURE IN SCOTLAND

In one important respect the pattern of coastal fishing activity in Scotland stands apart from the rest of the UK, for marine aquaculture plays a very important role in the country's coastal economy. With production of farmed salmon currently valued at around £265 million (*Fish Farming International*, May 2001), Scotland retains its position as the second largest producer after Norway. Moreover, many of the 6,500 jobs attributable to the aquaculture industry are located in the remoter rural areas of the Highlands and Islands, where some 80% of total production occurs. From its origins in Loch Ailort in 1969 salmon farming has grown rapidly - especially since the late 1980s - to reach almost 120,000 tonnes in 1998.

Aquaculture production is dispersed among several hundred sites around the northern and western coasts, but ownership of the industry is concentrated in a relatively small number of firms many of which are subsidiaries of multi-national food or chemical corporations. Improvements in cage design, feeding systems and pollution control have encouraged the adoption of further economies of scale so that much of the recent growth results from the intensification of production at existing locations rather than the proliferation of new sites. As a result, employment growth has failed to keep pace with the expansion in output, as the case of Marine Harvest McConnell demonstrates; between 1989 and 1997 production by the firm increased fourfold while employment grew by only 50% (Westbrook, 1999).

The dramatic growth of salmon aquaculture is in marked contrast to the fortunes of the wild salmon fisheries in Scotland and concerns have been expressed that the survival of wild salmon populations in some Scottish rivers may be under threat from the impacts of salmon farming (Nickson, 1997). In general the very low population of wild Atlantic salmon is attributed mainly to overfishing at sea and to the deteriorating conditions in salmon rivers throughout north west Europe. But the decline of wild salmon stocks in Scotland has also been linked to the siting, concentration and management of coastal fish farms in general and to the spread of sea lice infestations in farmed salmon - readily treatable through chemical biocides - to the unprotected wild salmon, in particular. Other concerns focus upon the risks of genetic interference resulting from the escape of genetically modified farmed salmon and the possible spread of diseases like the infectious salmon anaemia.

Wild salmon fishing in Scotland has experienced a dramatic turnaround over the past two decades: in the mid-1980s almost three quarters of all wild salmon were taken by commercial net fishermen operating on the coast or in the lower reaches of the estuaries and only a quarter by recreational rod and line fishermen, while a decade or so later the situation had been almost reversed with roughly two thirds of the catch taken by recreational fishermen and only a third by commercial fishermen. The net economic value of the rod fishery is estimated to be in the order of £350 million, thus far exceeding the first-hand value of production from salmon farming and approaching the first-hand sales value of Scotland's sea fisheries.

The administration of the salmon farming industry in Scotland is complex and the rapid expansion of the industry has largely frustrated the development of a fully coherent regulatory system. There are two sequential approaches to the system of regulation: the first concerns the location of the fish farms (planning controls) and the second refers to the setting of standards for and the monitoring of effluent discharges from the fish farms (water quality controls). The latter function has been vested in the Scottish Environmental Protection Agency with powers to issue discharge consents. The situation concerning planning control is more complicated. Planning decisions in Britain are normally the responsibility of the local authorities but, because their jurisdiction does not extend beyond the low water mark, the Crown Estate Commissioners as nominal owners of the seabed have assumed a voluntary planning role over access to the marine environment in the case of aquaculture development. As van der Schans (1999) has pointed out, the position of the Crown Estate has been subject to criticism as being neither accountable nor sufficiently impartial because the Estate stands to benefit financially from the granting of leases. Attempts to strengthen the planning process through the decision of the Scottish Office in 1988 to establish an independent but non-statutory committee to review the more contentious lease applications had little effect, leaving local communities with a sense that they were powerless to resist the 'colonisation' of their coastline by external interests wishing to establish fish farms in the area.

However, the situation in the Shetland Islands is very different. As a consequence of the *Zetland County Council Act 1974*, the Shetland Islands Council as harbour authority is able to assert control over development in its coastal waters. According to van der Schans (1999), the Council has used its special powers to introduce a restricted access policy and to create an aquaculture industry that is locally owned and integrated into the local economy. In essence, therefore, the Shetland Islands' communities have been more closely involved in decision making about the expansion of the aquaculture industry and, as a consequence, are much more at ease with this somewhat controversial aspect of modern fisheries development.

6. EMERGING ISSUES

Even though in England and Wales at least, inshore fisheries management enjoys the benefits of devolved and decentralised structures with strong stakeholder participation, there is concern among some inshore fishermen that their interests are not being adequately represented. In part this complaint refers to the method of selecting industry members of the SFCs through nominations by MAFF intentionally designed to avoid the risks of 'constituency politics' which might result from the direct representation of local associations or particular user groups. The system is, however, perceived by some in the industry as deliberately screening out those with strong political views.

Possibly of greater concern is the apparent decline in the representation of the inshore sector in the broader arena of fisheries politics in Britain, resulting from a progressive sidelining of local and regional associations which more strongly reflect the small boat interests. In England and Wales, the structural reorganisation of the National Federation of Fishermen's Organisations in 1995 reflects this tendency: direct representation of the local associations on the executive committee was replaced by nominations from the more robust, active and commercially oriented POs which often exclude under 10 m vessels from their membership, though inshore interests are still nominally reflected through the federation's regional committees. The small boat sector is perceived as lacking in political leadership, relatively powerless in negotiations and starved of quota in favour of the POs which now account for over 95% of all quota allocations.

The problem of representation for the inshore sector is in some ways less evident in Scotland where the executive of the Scottish Fishermen's Federation is made up of representatives of the main sectoral and regional associations, with the latter providing a limited means of access for inshore interests. Nonetheless, the inshore interests are dwarfed by the much stronger representation of the offshore industry, and there is an uneasy relationship between the federation and some of its constituent associations with strong small boat interests. Some regional associations, particularly from the small boat dominated west coast districts, remain through choice outwith the federal structure.

Recent legislation, combined with a strong groundswell of opinion in favour of adopting a precautionary, ecosystem based approach to fisheries management, suggests a potential for radically altering the established systems and structures for inshore fisheries management. Some of these new approaches underline the pressure to review the current arrangements and to establish closer integration of fisheries and marine conservation management especially within the inshore zone, with financial provisions to encourage a switch from the more abusive fishing practices to methods which minimise the risk of damage to the marine ecosystems and their habitats. Questions are inevitably raised over the competence of SFCs, as presently structured, to oversee the implementation of the necessary policy shift and over the suitability of their existing powers. Certainly the slow and cumbersome byelaw making process seems ill suited to meet this particular challenge. In this context, it will be interesting to chart the development of a further 'greening' of inshore fisheries management north and south of the border where wholly different structures are in place.

By comparison with the level of scientific input into stock assessment exercises for quota regulated demersal and pelagic stocks - many of which are in a critical state - scientific investment in inshore fisheries is much less, despite the research programmes being conducted into key shellfish species by both the Scottish and English laboratories. Moreover, the recording of landings for many inshore species is poor and official statistics are probably subject to gross underestimation. It is important, therefore, to extend data collection and analysis at the local level. In England and Wales many of the larger SFCs now have a limited capacity for scientific research and investigations have already been initiated as for example in relation to the mussel fisheries in the Wash (Eastern SFC) and lobsters in the North Eastern Sea Fisheries District. Equally vital is improving the flows of detailed catch information from the industry to the management organisations so that the status of the inshore stocks can be more reliably assessed and their management put on a sounder footing.

7. CONCLUSION

Tension between the central state and the concept of local governance is most likely to arise in the case of inshore fisheries management where there is a *prima facie* case for arguing that the harnessing of local knowledge and a sense of co-responsibility provides the firmest basis for effective management. Clearly in Britain the administrations have sought to manage this tension in very different ways, though in both instances it can be argued that the centre still exercises considerable authority.

In the case of Scotland, central control is much stronger and more transparent. There has only been a very limited devolution of responsibility to industry based organisations and a correspondingly weak level of decentralisation. At present a more widespread deployment of Regulating Orders would seem to offer the best prospect for the development of local governance, though doubts exist about the legal competence of local management committees to enforce the management measures. Elsewhere the progress of several other applications is mired in local politics.

By contrast there exists a nominally fully devolved and decentralised system of management embracing the inshore fisheries of England and Wales. Significantly, the constitution of the SFCs effectively extends the competence of local authorities beyond the conventional limits of the low water mark to the present boundary of the 6 mile zone. Beyond that point and out to the 12 mile limits management responsibility rests with central government. But even within the inner zone, MAFF manages to retain a strong element of influence and steering over the SFCs. It can exert this influence in two ways: first through the selection of appointed members to the committees and, secondly, through the process of byelaw making in which MAFF retains the ultimate decision making role. In future, should a decision be made to undertake a measure of central funding of SFCs, its influence will be further strengthened. At present it seems that MAFF is content to allow SFCs to continue in their role of reactive management, while retaining the initiative for proactive decision making. Although central government would no doubt have considerable difficulty in rescinding the rights and responsibilities already vested in SFCs over the years, there is no doubt that MAFF remains in a position to curb their influence, should it choose to do so.

8. REFERENCES

Addison, J. (2000) National and regional management of lobster fisheries in the UK, in O. Tully (ed.), *Management of Irish Lobster Fisheries: A Discussion with Industry*, Trinity College, Dublin, pp. 42-49.

Association of Sea Fisheries Committees (2000) *2000 and Beyond. Report on the Role of Sea Fisheries Committees on the Management and Regulation of the Coastal Waters of England and Wales*, Association of Sea Fisheries Committees, Malton.

Cameron Report (1970) *Regulation of Scottish Inshore Fisheries*, Report of the Scottish Inshore Fisheries Committee Presented to Parliament by the Secretary of State for Scotland, Her Majesty's Stationary Office, Edinburgh.

Gray, M.J. (1995) *The Coastal Fisheries of England and Wales, Part III: A Review of their Status 1992-1994*, Fisheries Research Technical Report No. 100, Directorate of Fisheries Research, Lowestoft.

Harrison, N. (1993) *RSPB Submission to the Agriculture Committee on the UK Sea Fish Industry*, Royal Society for the Protection of Birds, Sandy.

House of Commons (1999) *Sea Fishing, Eighth Report of the Agriculture Committee: Column 1, Session 1998-99*, The Stationery Office, London.

Nickson, Lord (1997) *Report of the Scottish Salmon Strategy Task Force*, The Scottish Office, Edinburgh.

Phillipson, J. (1999) The fish producers' organisations of the UK: a strategic analysis, in D. Symes (ed.), *Alternative Management Systems for Fisheries*, Blackwell Science, Oxford, pp 79-92.

Scottish Executive (2001) *Working Together for Scotland: A Programme for Government*, Scottish Executive, Edinburgh.

Scottish Office (1996) *Scotland's Coasts: A Discussion Paper*, Scottish Office, Edinburgh.

Seafish (1991) *The Highland Region - The Prospects for Fisheries and Aquaculture in the 1990s*, Consultancy Report 41, Seafish Technology, Sea Fish Industry Authority, Edinburgh.

Symes, D. and Phillipson, J. (1998) Fishing within limits: inshore fisheries management and the concept of local preference, in D. Symes (ed.), *Property Rights and Regulatory Systems in Fisheries*, Blackwell Science, Oxford, pp. 201-215.

Van der Schans, J.W. (1999) Governing aquaculture: dynamics and diversity in introducing salmon farming in Scotland, in J. Kooiman, M. van Vliet and S. Jentoft (eds.), *Creative Governance: Opportunities for Fisheries in Europe*, Ashgate, Aldershot, pp. 95-118.

Westbrook, S. (1999) *The Economic Impact of the Operations of Marine Harvest McConnel in the Highlands of Scotland*, Marine Harvest McConnell, Edinburgh.

Chapter 7

IRELAND

Nathalie Steins

1. INTRODUCTION

The Irish Republic is essentially an island state. Its long, irregular and in parts deeply incised coastline shelters nearly 900 ports, small harbours and landing places - though many lack the basic infrastructure for handling regular consignments of fish. At the time of entry into the EU in 1973, Ireland's fisheries were seriously undervalued, weakly developed and poorly organised. Landings totalled a mere 85,500 tonnes. Over the past quarter of a century, while some Member States have faced declining landings and increasing pressures to reduce fishing capacity, Ireland has experienced a period of new investment, vigorous modernisation and strong growth in fishing activity. By 1997, landings had more than tripled to *circa* 300,000 tonnes. Nonetheless, fishing still accounted for no more than 1% of GDP and employment.

Inshore fisheries make an important contribution to the overall fishing economy and to local employment. They are estimated to earn *circa* £IR30 million for the Irish economy and to provide a total of 3,700 seagoing jobs (Bord Iascaigh Mhara, 1999). Up to 70% of the Irish fleet depends on fishing within the 12 nm zone for the exploitation of quota and non-quota species. A wide range of vessels can be found within this zone, including not only the typical, small inshore boats but also the much larger vessels of the Irish pelagic fleet. Moreover, several EU Member States - including the UK, France, Belgium and the Netherlands - have historic fishing rights in the 6-12 nm zone.

Apart from the legal definition of the 12 nm zone, there is no formal designation of inshore fisheries in Ireland. A recent report by the Irish Sea Fisheries Board (Bord Iascaigh Mhara, 1999) makes the relatively

D. Symes and J. Phillipson (eds.), Inshore Fisheries Management, 119–137.

'generous' definition of inshore vessels as under 15 m, though such vessels clearly have the capacity to range well outside the 12 nm zone. A further subdivision of the inshore fleet, separating out the truly coastal operations of seasonal, part time and full time fishermen who exploit both quota and non-quota species exclusively within the 0-6 nm zone mainly with vessels under 12 m, would therefore be useful.

2. IRELAND'S INSHORE FISHERIES

One of the problems confronting any attempt to develop a comprehensive management strategy for inshore fisheries in Ireland - as in most European countries - is the huge variety of species and production methods which are encountered in inshore waters. The picture is complicated still further by the particular permutations of target species and therefore fishing gears according to season and geographical location. Nonetheless, inshore fisheries in Ireland can be divided into two broad categories - capture fisheries which include the increasingly important sea angling and culture fisheries for both finfish and shellfish - and each of these broad divisions is capable of further subdivision. Capture fisheries are dealt with in the present section while the expanding aquacultural sector is considered separately.

2.1 Production

The principal capture fisheries include both demersal (mainly cod, haddock and whiting, but also plaice and sole) and pelagic (mainly herring) species, together with the anadramous salmon, decapods (*nephrops*, lobster, crab and crayfish) and bivalves (oysters and mussels). All except the bivalves can be exploited outside the 12 nm zone as well as inside it. Fishing methods deployed within the inshore zone, including both static and mobile gears, vary according to species (see Table 7.1), season, location and vessel size. In general, mobile gears are used principally in the prosecution of demersal and pelagic fisheries and mainly by vessels over 10 m. The exceptions are the use of draftnets for salmon in estuarine locations and, more generally, dredges for a range of shellfish in the relatively protected waters of the shallower bays around the western and southern coasts. Conversely, the smaller the vessel the more likely it is to use static gear - notably the pots and creels deployed seasonally by boats under 10 m in the lobster, crab and crayfish fisheries around the Irish coast.

In 1997, Irish inshore vessel landings, excluding aquaculture, were valued at £IR21.9 million or 16.4% of the total value of the capture fisheries (see Table 7.2). The largest single contribution came from the decapod fisheries and, in particular, from lobsters (£IR4.3 million) which alone accounted for almost 20% of the overall value of the inshore fisheries.

Table 7.1 Ireland: fishing methods and inshore fisheries

Mobile gears		Static gears	
Demersal otter trawls	most whitefish species	*Gillnets*	cod, pollack, saithe, ling
Midwater pair trawls	herring, sprat	*Tangle nets*	monkfish, turbot, crayfish
Beam trawls	sole, plaice	*Pots and creels*	lobster, brown crab, spider crab, velvet crab, shrimp, welk
Seine nets	cod, haddock, whiting	*Longlines*	conger eel
Draftnets (estuaries)	salmon	*Jigging*	mackerel, squid
Dredges	scallops, oysters, mussels, clams		

Source: Bord Iascaigh Mhara, 1999

Table 7.2 Ireland: landings by vessels under 15 m (excluding aquaculture) 1997 by volume and value

	Volume	Value	
	(tonnes)	£IR thousand	%
Demersal	5,000	6,500	29.6
Salmon	600	1,500	6.8
Shellfish of which	8,823	13,990	63.5
(a) *Decapods*	*1,939*	*7,991*	*36.3*
lobster	500	4,349	
crab	1,026	1,489	
shrimp	359	1,673	
crawfish	24	330	
nephrops	30	150	
(b) *Bivalves*	*932*	*2,664*	*12.1*
oyster	750	2,171	
scallop	80	240	
palourdes	10	128	
razorshell	28	84	
cockle	64	41	
(c) *Others*	*5,952*	*3,335*	*15.2*
whelk	2,800	1,232	
periwinkle	3,152	2,103	
Total	14,423	21,990	100

Source: Bord Iascaigh Mhara, 1999

Table 7.3 The Irish inshore fleet, 1997

		Inshore vessels (length in metres)					Employment	
		< 10	10-13	13-15	Total	%	Total	%
Eastern	(no)	224	82	18	324	18	706	20
	(%)	69	25	6	100	-	-	-
South West	(no)	457	170	16	643	35	1,184	33
	(%)	71	26	3	100	-	-	-
West	(no)	517	115	9	642	35	1,168	33
	(%)	81	18	1	100	-	-	-
North West	(no)	114	107	15	236	12	506	14
	(%)	48	45	6	100	-	-	-
Total	(no)	1,312	474	58	1,845	100	3,564	100
	(%)	71	26	3	-	100	-	-

Source: Bord Iascaigh Mhara, 1999

2.2 The fishing fleet

The registered inshore fleet comprises a total of nearly 1,850 vessels under 15 m in length, with a marked geographical concentration in the western and south western counties, especially Galway (455) and Cork (393). Roughly another 900 unregistered inshore vessels need to be added to make up the totality of the active inshore fleet. As Table 7.3 and Figure 7.1a demonstrate, the size structure of the registered fleet is clearly skewed towards the smaller vessels under 10 m: some 70% of all inshore vessels are under 10 m and only 3% are in the largest inshore category (13-15 m). The dominance of these small coastal boats is greatest in the west of Ireland and, paradoxically, rather less marked in the north west (Donegal) where less than half the inshore vessels are under 10 m.

Although Figure 7.1b clearly shows the evidence of high levels of investment in the expansion and renewal of the inshore fleet following entry into the European Community in the early 1970s - with 46% of the existing fleet built in a ten year period from 1975 to 1985 - Ireland's inshore fleet is an ageing one, especially when compared with the offshore sector. A significant proportion is over 35 years old, while only 12% have been built in the past fifteen years.

2.3 Employment

Most of the vessels engaged in the inshore sector are skipper owned and rarely employ more than one or two men. Although many crew members will share family ties with the skipper of a particular boat, some may fish with more than one skipper during the year depending upon the combination of fishing seasons.

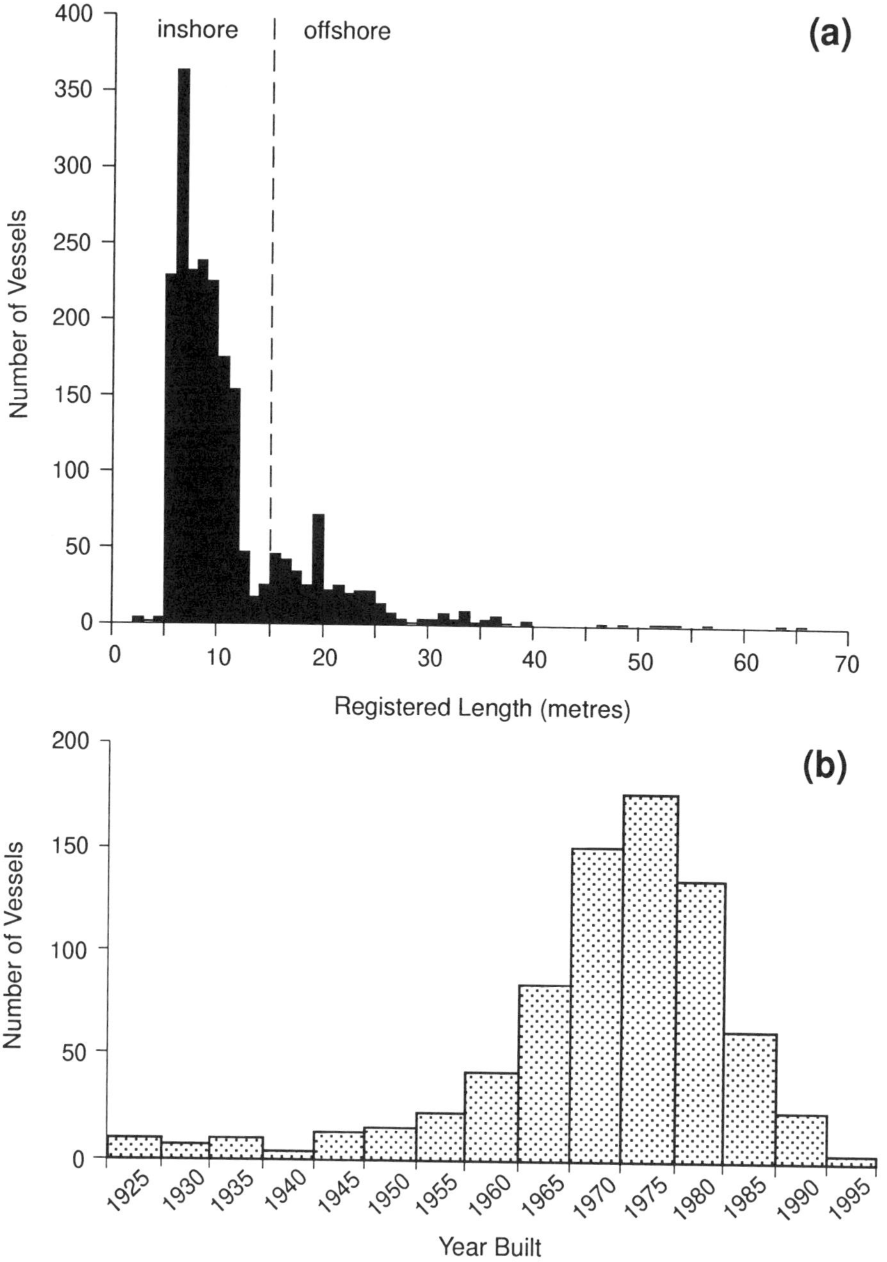

Figure 7.1 Ireland: The Fishing Fleet, 1998
(a) Vessel Size (b) Date of Construction
(Source: Bord Iascaigh Mhara, 1999)

Table 7.4 Ireland's fisheries dependent regions, 1991

	Total			Fishing related			
	Population	Employment	GRP	Employment		GRP	
	(thousand)	(thousand)	ECU million	No	%	ECU million	%
Donegal	130	33	1,021	3,266	9.9	61	6.0
Galway and Mayo	63	22	514	1,144	6.4	20	3.0
Cork and Kerry	107	32	1,006	1,990	6.2	37	3.7
Waterford and Wexford	88	26	754	1,286	4.9	21	2.8
Eastern	170	51	1,387	1,502	2.9	25	1.8
Dispersed	50	18	n.a	1,340	8.9	14	n.a
Ireland	3,540	1,090	9,450	10,527	1.0	178	0.5

Source: Commission of the European Communities, 1992

Overall the inshore sector is estimated to provide *circa* 3,700 jobs, of which roughly 2,700 are full time, thus accounting for around half of all onboard employment in Ireland's fishing industry (Bord Iascaigh Mhara, 1999). The employment multiplier - creating jobs onshore - is likely to be quite low reflecting the generally low volumes (but high unit values) of landings associated with the small inshore vessels. Again, a significant proportion of the jobs in fish processing and other fishing related activities ashore is likely to be part time and/or seasonal.

Geographically the distribution of the inshore fleet, the employment arising and the level of economic deprivation are clearly correlated. Employment is again highest in the south western and western districts (see Table 7.3) which, together with the north west, represent the areas of relative economic disadvantage. These areas also coincide with the largest of the fisheries dependent regions (see Figure 7.2) identified in the 1991 Regional Socio-Economic Studies (Commission of the European Communities, 1992). The relative importance of the fishing industry to both employment and gross regional product at the start of the 1990s is indicated in Table 7.4.

Inshore fishing is a strongly seasonal activity with peaks in the summer months (June, July and August), coinciding with the salmon, lobster and crab fisheries, and is at its lowest ebb in the winter months (January and February). The number of days fishing will vary markedly according to vessel size, patterns of fishing activity and exposure to bad weather. It is likely that for a majority of inshore fishermen, including some who are

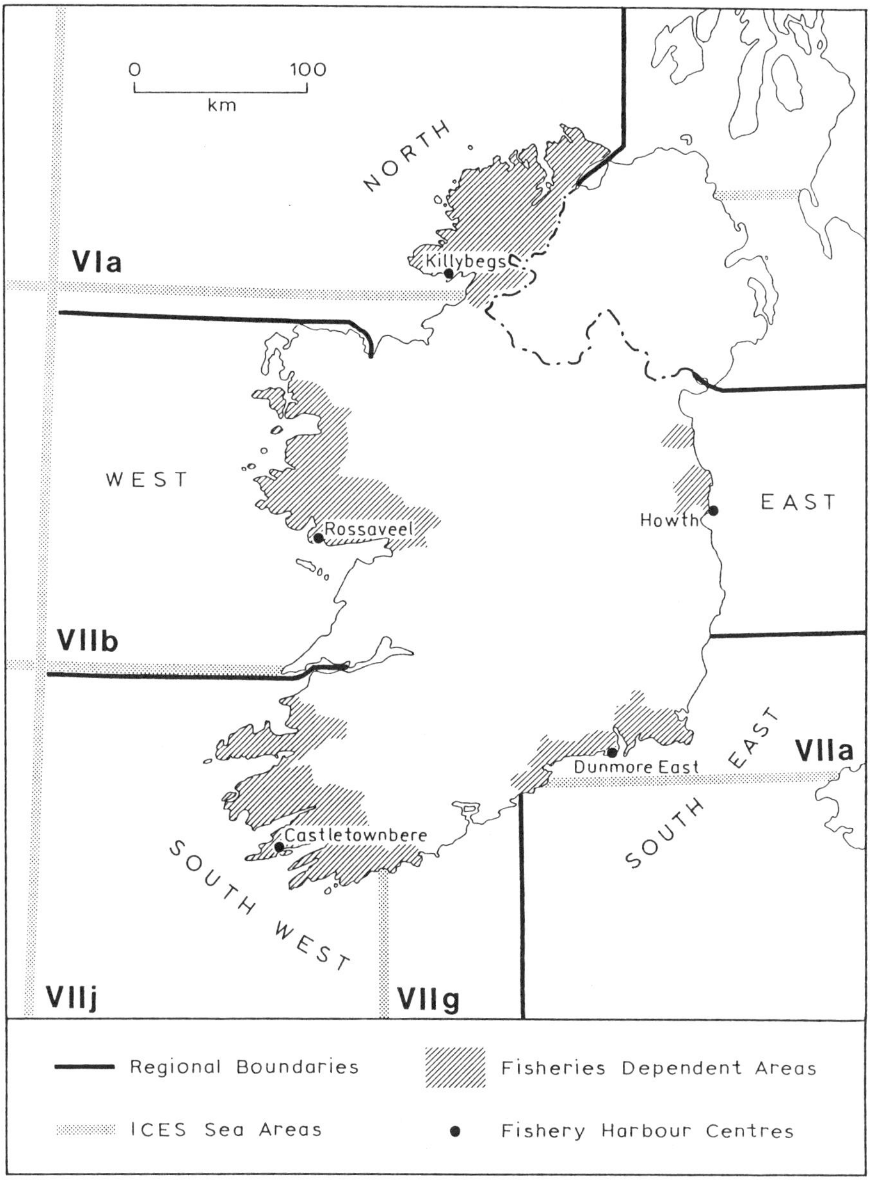

Figure 7.2 Ireland: Fishing Regions and Fishing Dependent Areas

active all year round, fishing will fail to yield a sufficient living wage for the fishermen and their families. In the western coastal districts, in particular, where there are relatively few sources of alternative employment - especially in the winter months - there is likely to be a relatively high dependence on social welfare payments to supplement incomes.

Age profiles for employment in the inshore and offshore sector (Figure 7.3) reveal some quite pronounced differences. These may, in part, simply reflect the comparatively strong growth of the offshore sectors in recent years. But the inshore fishermen are noticeably older. The differences are most marked between skippers where the median age in the inshore fleet is around 40 compared with 35 on the larger offshore vessels. A weak level of recruitment of new entrants into the inshore sector is indicated and this is matched by a tendency for inshore skippers to remain in active employment - especially on a part time or seasonal basis - until much later in life.

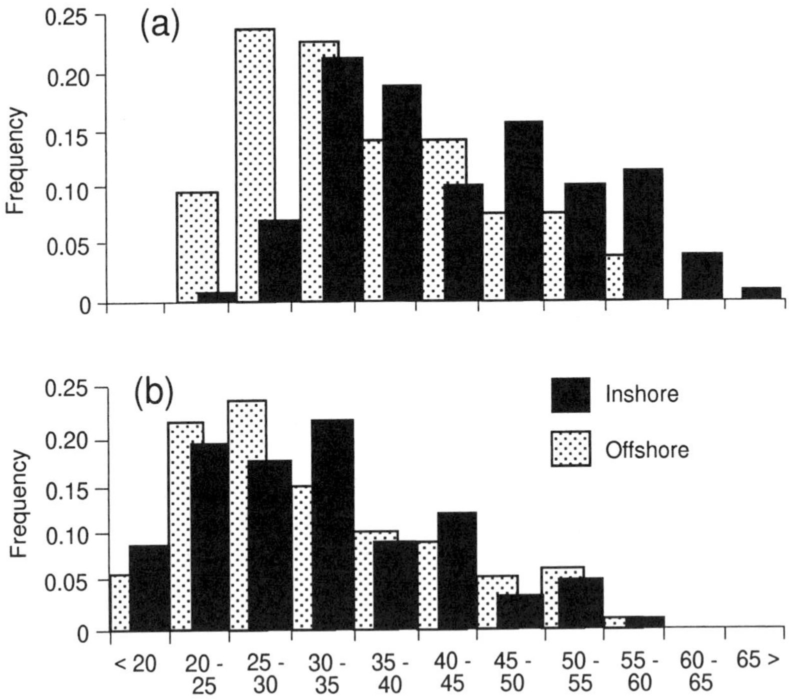

Figure 7.3 Ireland: Age Profiles (a) Skippers; (b) Crew Members (Source: Bord Iascaigh Mhara, 1999)

3. MANAGEMENT SYSTEMS

3.1 The legislative basis

Although systems of sea fisheries management in Ireland are now subordinate to the provisions of the Common Fisheries Policy, the *Fisheries (Consolidation) Act 1959* remains the basic legislative instrument. Part XIII of the 1959 Act, as amended in 1962, 1978, 1983 and most recently in 1994, provides for the control of access to fishing grounds, the registration and licensing of fishing boats, the establishment of conservation measures and the monitoring and enforcement of policy in respect of both the inshore and offshore sectors. All vessels over 5 m have to be registered and licensed for the purpose of commercial fishing.

In part, inshore fisheries are considered in the provisions of the *Fisheries Act 1980* establishing Fisheries Boards with responsibility for the management, conservation, development and promotion of inland fisheries and sea angling, together with the protection - but not development - of the molluscan shellfisheries. Conditions governing the issue of salmon net licences by the regional Fisheries Boards are laid down in the *Control of Fishing for Salmon Orders* of 1980 and 1982. These provide the basis for fixing the maximum number of licences for the driftnet (currently 847) and draftnet (604) fisheries. Salmon fishing within the Shannon estuary is controlled separately through the *Shannon Fisheries Act 1935*. Otherwise, there is no specific national legislation referring to inshore fisheries management in Ireland, apart from the special case of the Lough Foyle fisheries in north west Ireland, shared between the Republic and Northern Ireland, which are subject to a local management system under the terms of the *Foyle Fisheries Act 1952*.

3.2 The institutional framework

The management structures for inshore fisheries in Ireland follow broadly the pattern of a 'command and control' system with few concessions to devolved or decentralised approaches (Figure 7.4). Overall responsibility for the regulation and development of fishery resources rests with the Department of the Marine and Natural Resources. In terms of research and advice, the work of the Department is supported by two state funded agencies: the Irish Sea Fisheries Board (Bord Iascaigh Mhara or BIM) with responsibility for technical, financial and marketing advice and for training in both the capture and culture fisheries; and the Marine Institute established in 1991 to support and promote the sustainable development of Ireland's

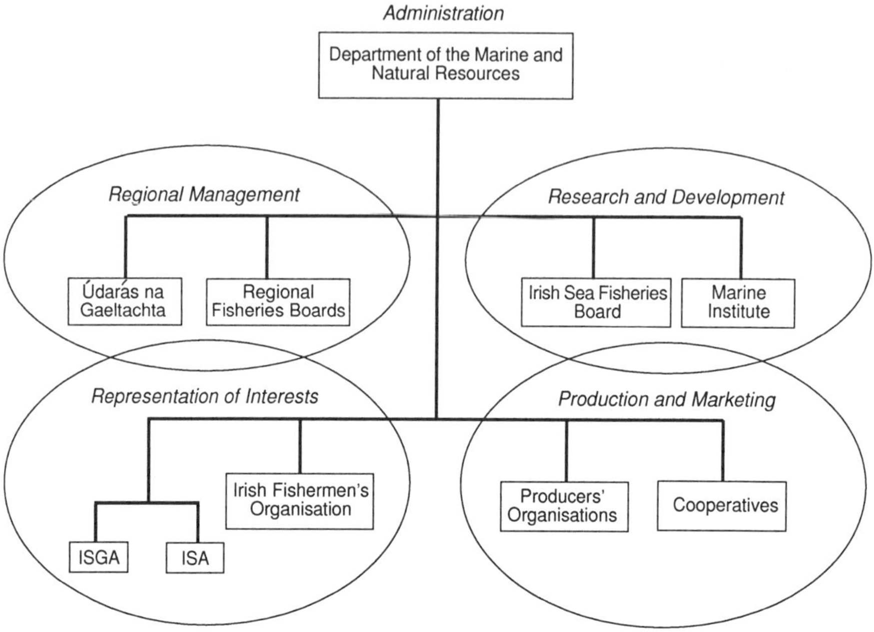

Figure 7.4 Ireland: Structure of Fisheries Management

marine fisheries through scientific research, organised according to eight 'sectoral' areas - though none is specifically concerned with inshore fisheries *per se*. The Institute was also responsible for implementing Ireland's *Operational Programme for Fisheries, 1994-1999.*

A third state agency, *Údarás na Gaeltachta*, with responsibility for the development of the Gaeltacht - Irish speaking districts found largely in the more remote north western (Donegal), western (Mayo and Galway) and south western (Kerry and West Cork) coastal districts - is involved in the development of inshore fisheries, mainly with reference to aquaculture, through support for technical research, grant aid and training provision. Recently Údarás has also become involved in the capture fisheries including the development of salmon rivers and the enhancement of lobster and scallop fisheries within the Gaeltacht.

Otherwise, the concept of decentralised and devolved management for Ireland's inshore fisheries rests with the regional Fisheries Boards established under the *Fisheries Act 1980*. The seven regional Boards, concerned primarily with inland fisheries management and with only a limited remit in relation to coastal fisheries out to the 12 nm limit, are made up of both elected and nominated members. A central Board, comprising the chairs of the regional Boards together with Ministerial appointees including the chairperson, oversees and coordinates the work of the seven regional Boards.

Ireland's three producers' organisations (the Irish FPO, the Killybegs FO and the Irish South and West PO) also exert a considerable influence on the day to day management of the quota regulated demersal and pelagic species: their memberships include significant numbers of <15 m vessel owners but relatively few in the <10 m class. A separate producers' organisation has been established to handle the sales of farmed salmon.

The cooperative movement in Ireland has a long, if somewhat troubled, history of providing a marketing structure for small enterprises in the primary sectors of agriculture and fishing. Today there are a total of 17 local or regional fishing cooperatives concerned with organising the sales of fish and providing a basic range of services and facilities for their members. They have been instrumental in the recent strong growth of the lobster fisheries and largely responsible for implementing the lobster v-notching programme with the aid of PESCA funding. A few of the cooperatives are also engaged in processing activities.

Inshore fishermen rely mainly on the Irish Fishermen's Organisation (IFO), formed in 1974, to represent their interests in consultations and negotiations with the higher echelons of the management structure - though only a relatively small proportion are members of the IFO. In recent years specialised associations have been established to promote the particular interests of two of the most dynamic sectors of the Irish inshore industry - the Irish Salmon Growers' Association (ISGA) and the Irish Shellfish Association (ISA). Both are affiliated to the more powerful Irish Farmers' Association and both have been instrumental in developing management strategies for their particular sectors.

3.3 Management systems

It seems reasonable to argue that Ireland's inshore fisheries are generally undermanaged - a view shared by the inshore fishermen themselves (Bord Iascaigh Mhara, 1999). But opinion is divided as to how best to remedy the situation.

Management regimes are in place for all quota species, although the inshore sector accounts for only a small share of the total landings of the main demersal species - cod, haddock, whiting, plaice and sole - which are managed through the producers' organisations. But for herring - the principal quota species caught within the 12 nm zone, though mainly by offshore vessels - two distinct management regimes have been developed. In the

Celtic Sea (ICES VII g and j) the fishing lasts from October to February: the TAC is divided into weekly quotas and allocated according to vessel size. The fishery is closed each week for 48 hours (Saturday and Sunday) and landings can only be made at designated ports (Molloy *et al.*, 1996). The western fishery (VIa and VIIb), centred on Killybegs in County Donegal, is now managed centrally along broadly similar lines but until 1996 was in the hands of local committees made up mainly of industry representatives. Failure of the local management system was due largely to their inability to limit entry to the fishery and to agree an equitable vessel quota system, together with a lack of legislative support for enforcing the regulations.

Apart from herring, the most closely regulated inshore fishery is the wild salmon. The salmon fisheries have been subject to almost continuous scrutiny through reviews and commissions seeking to strike an appropriate balance between the commercial coastal fishery and the recreational inland fishery. The trend has been for a progressive shift towards recognising the greater economic value of the latter. Currently the coastal driftnet fishery is confined to a short summer season lasting through June and July. Opinions differ as to how best to promote the freshwater fishery, whether through an outright ban on drift and draftnet fishing or through the introduction of much tighter controls on the coastal fishery.

The Salmon Management Task Force which reported in 1996 clearly did not envisage an end to the coastal fishery arguing that 'an equitable sharing between commercial and recreational sectors is not only possible, but desirable' (Salmon Management Task Force, 1996: p. 14). Among the Task Force's recommendations were a cap on commercial licences, a reduction in the area of driftnetting from 12 nm to 6 nm offshore, postponement of the opening dates for the drift and draftnet fishing seasons, the introduction of a four day fishing week and tighter regulation of the daily opening and closing times for the driftnet fishery. A new management regime was proposed for the 1997-2000 period based on creating optimal spawning populations, improving the escapement levels for spawning salmon, the introduction of quotas and the formulation of local catchment management plans. The development of the wild salmon fisheries was to be overseen by a National Commission and implemented through local management committees. To date few of these recommendations have materialised.

Management regimes for other non-quota species are, to all intents and purposes, absent. In a few instances, local cooperatives have taken the initiative of introducing local regulations for the control of particular fisheries, as with the introduction of a Fishery Order at the request of the Valentia Harbour Cooperative in Kerry for the regulation of the scallop fishery and the draft order sought by the South West Fisheries Cooperative for the management of crab and lobster stocks by introducing area and seasonal closures and attaching more restrictive conditions to shellfish licences (Bord Iascaigh Mhara, 1999). Perhaps the most important self-management action undertaken in the shellfish sector, however, has been the introduction of a nation wide v-notching programme to protect berried lobsters, inaugurated by the Irish Lobster Association with the assistance of BIM, in part funded through PESCA and carried out through a large number of participating cooperatives. The extension of the scheme to include crayfish is now under consideration.

4. AQUACULTURE

4.1 Production

As in many other countries bordering the North Atlantic, the past two decades in Ireland have witnessed a remarkable expansion in the production of fish and shellfish through aquaculture. Still in its developmental stage, however, Ireland's output still lags well behind its leading European competitors - Norway and Scotland - and accounts for somewhat less than 5% of world production. Development has been concentrated mainly along the western seaboard from Donegal in the north to West Cork in the south.

There are two very distinctive elements to Ireland's aquaculture industry. The dominant finfish production of Atlantic salmon - and, to a much lesser extent, sea trout and turbot - located almost exclusively in these western districts, remains largely in the hands of a relatively small number of national or foreign owned companies. The Scottish Gaelic Seafoods, for example, controls nearly 50% of Ireland's farmed salmon production. In less than 20 years, total output in the finfish aquaculture sector had risen from 600 tonnes in 1980 at a value of IR£894,000 to around 17,000 tonnes in 1998 at value of IR£47.5 million. By the mid-1990s there were some 33 finfish farms employing around 1,100 persons, of whom 600 were full time.

By contrast, aquaculture development in the shellfish sector is both less dramatic in scale and mainly the result of small scale local initiatives, often organised on a cooperative basis. Two species dominate present production: the bottom and rope cultures of mussels (*Mytilus edulis*) originating in the south around Waterford, Youghal and Bantry Bay but subsequently diffusing throughout much of Ireland's coastline, have grown from some 4,700 tonnes in 1980 to around 18,000 tonnes in 1997; while the cultivation of Pacific oysters (*Crassotrea gigas*) has expanded from less than 50 tonnes in the early 1980s to almost 3,400 tonnes in 1997. Production of other shellfish, including the native oyster, clams, scallops and abalone, is less than a thousand tonnes. Altogether the value of shellfish aquaculture production in

1998 was slightly over £IR13 million (23,200 tonnes) or roughly a quarter of the revenue arising from the shellfish sector.

4.2 Management

Aquaculture offers several key advantages in the quest to expand the fisheries sector:

- it uses existing resources;
- it supplements the depleted natural fish stocks;
- it is free from quota restrictions imposed through the CFP;
- it can be stimulated through the application of science and technology;
- it provides employment in areas where job opportunities are scarce; and
- it generates substantial export earnings.

Partly for these reasons the Irish government has striven to minimise the level of interference in its development, especially in the case of salmon farming where largely foreign capital investment has been involved. Whereas traditional forms of shellfish aquaculture were covered by the provisions of the *Fisheries (Consolidation) Act 1959*, salmon farming was less closely regulated, despite its greater potential to cause conflicts of interests with existing inshore fisheries and to arouse public opposition locally. Indeed, with the express purpose of simplifying and thereby accelerating the processes involved in licensing new salmon production units, 'designated areas' were introduced through the *Fisheries (Amendment) Act 1980* wherein the Minister had the sole responsibility for determining the outcome of a licence application on the basis of consultation with 'relevant bodies' rather than the more long drawn out system of public participation implicit in public enquiries (Steins, 1998). However, in the face of challenges from the High Court in the early 1990s, the government was forced to introduce the *Fisheries (Amendment) Act 1997* establishing an Aquaculture Licence Appeals Board. In affect this removed direct responsibility for licensing from the Minister and re-established the rights of full public consultation in the hopes of reducing the level of conflict and public opposition over the licensing of new salmon farms - albeit at the cost of delaying the decision making process.

5. MANAGEMENT ISSUES

Many of the key issues have already been alluded to in the preceding account of the inshore fisheries and their management, most notably the conflicts between aquaculture and capture fisheries in the coastal zone and the allocation of fishing rights for the increasingly scarce wild salmon resources. Some further elaboration of these issues is warranted along with matters relating to the economic viability of the inshore sector and future recruitment to the industry.

Although aquaculture has rightly been recognised as the most dynamic sector of Ireland's fish producing industry and one which has made important contributions to the coastal economies in western Ireland in particular, its future development is fraught with problems. The potential for further growth is constrained by a number of factors:

- a lack of suitable production sites as recent expansion has already led to 'overcrowding' in the more favoured locations;
- increasing threats from diseases and growing concern over the use of chemicals in their treatment especially in the case of finfish culture;
- saturation of the markets for certain farmed fish products and a consequent fall in profit margins;
- the prospect that future investment will be curtailed following the loss of Objective 1 status for Ireland in the reform of the EU's Structural Funds and the ending of the PESCA initiative; and
- a growing tendency to question the real impacts of aquaculture especially in areas which rely heavily on the imagery of 'unspoilt natural beauty' to sustain the role of tourism as a principal source of income generation.

With the introduction of new planning procedures it seems likely that applications to develop new production sites - especially where promoted by outside interests - will face opposition from an alliance of local interests including inshore fishermen and those engaged in the tourism economy. Future expansion of aquaculture could well depend on the intensification of production among existing enterprises rather than the proliferation of new sites - but this is not without its own problems. Clearly, therefore, one of the issues confronting regional economic planners and those responsible for coastal zone management will be to find an appropriate balance between the conflicting interests of aquaculture, inshore fisheries and tourism.

A question of balance is also at the forefront of questions relating to the protection and exploitation of wild salmon stocks which presently support a small, intensely seasonal but nonetheless lucrative coastal and estuarine fishery and, at the same time, form the basis of an important inland recreational fishery with strong links to tourism. The issue is how to make the best socio-economic use of what has become a declining resource due to the physical disruption of river regimes, pollution and overfishing. Throughout the North Atlantic there has been a clear trend in the pattern of exploitation away from the high seas and coastal waters to a concentration on freshwater fisheries which are easier to control and economically more profitable (Fingleton and Whelan, 1994). Although the Salmon Management Task Force (1996) argued in favour of an equitable allocation of fishing rights between coastal and inland fisheries, the debate continues. Angling interests, supported by the tourism lobby, wish to see a phasing out of a coastal driftnetting - in line with several other Member States - or at least a moratorium on commercial fishing to allow stocks to recover. Coastal fishermen, not surprisingly, protest that the balance has already swung too far and that existing restrictions on fishing effort, which take no account of uncertain weather conditions, cause severe operational difficulties in both catching and marketing the salmon.

Efficient marketing is a general problem for inshore fisheries in Ireland. Globalisation of the food economy and the growing penetration of local, regional and national markets by imported produce is as much a problem for fisheries as it is for agriculture. Moreover, inshore fishermen face increasing competition on their own doorsteps from escalating production of farmed fish and shellfish. The effect is in some instances to cause a downward spiral of prices, profit margins and reinvestment in the inshore sector. These problems are exacerbated - especially in Western Ireland - by the region's peripheral location which adds significantly to the time and cost involved in getting quality fish to Europe's urban markets and by the underdeveloped infrastructure of the fishing industry in terms of the age of the fleet and the lack of modern handling facilities ashore. The designation of five state owned Fishery Harbour Centres at regionally strategic locations in Killybegs, Rossaveal, Castletownbere, Dunmore East and Howth (see Figure 7.2) still leaves many inshore fishermen at a serious disadvantage in gaining access to competitive markets able to pay top prices for quality produce.

Despite considerable investment following entry to the EC, Ireland's fishing fleet suffers from a relatively high age structure, with an average age of 30 and a significant proportion of vessels built before 1960. These older vessels are less able to cope with conditions encountered in offshore waters and there are increasing concerns for vessel safety. As a consequence a number of older boats over 15 m, including some beam trawlers, now concentrate their activities within the 12 nm zone giving rise to a growing incidence of gear conflicts with inshore boats deploying static gears. Many smaller inshore boats are ageing, in need of more frequent repair and lacking modern fishing and fish handling equipment, thus adding to the costs of fishing.

The combination of rising production costs and falling quayside prices has meant that some inshore fishermen have changed their activity patterns from those of professional full time or main time fishermen to behaviours more akin to 'hobby fishing' intended to maintain the family tradition and provide an income supplement but not a means of livelihood. For some, fishing is now perceived as an unrewarding career; young people are advised to look elsewhere for employment and the social reproduction of inshore fishing is thus threatened. This erosion of the viability and status of inshore fishing is more likely to occur in the more remote rural communities - precisely those areas where fishing is expected to perform a strong socio-economic function.

6. CONCLUSIONS

Since Ireland's accession to the EC in 1973, the fishing industry has generally experienced a period of expansion, modernisation and structural rationalisation. Its importance to the national economy remains slight, but for many of the coastal districts especially along the western Atlantic coastline, the fishing industry forms an integral and important part of the local economy. However, reform of the management structures and systems has failed to keep pace with the structural, economic and technical changes taking place within the fishing industry itself.

Certain sectors of the industry - not least the traditional small boat inshore fisheries - feel they have been neglected by government and they sense an uncertain future. Without a more coordinated approach to the management of the inshore sector, there are clear limits to the scope of local initiatives designed to protect and enhance fishing opportunities. Moreover, the persistence of a centralised approach to inshore fisheries management becomes a constraint on further development since decisions are less likely to be based on local experience and needs and will appear to be guided more by national political interests. As a result inshore fishermen feel dispossessed of the right to be involved in shaping their own futures. Stronger representation of fishermen's interests in the design, implementation and evaluation of local regulations is an essential ingredient of the protection of the local resource base. To a limited degree, Ireland's inshore fishermen have already revealed capabilities for positive action in improving the conditions of local fisheries through, for example, the lobster v-notching programme. They now need to demonstrate their willingness to take responsibility for their own futures on a much broader front.

Among the recommendations from the recent review of the inshore fisheries sector (Bord Iascaigh Mhara, 1999), several were concerned with institutional reform. These included:

- the establishment of an Inshore Fisheries Advisory Committee, drawn mainly from existing government organisations and the fishing industry, to advise the Minister on policy and strategies for the inshore sector;
- the development of a pilot scheme for the creation of local inshore fisheries development committees;
- the need to examine the feasibility of devolving certain aspects of inshore fisheries management to local groups;
- consideration of the possible location of inshore waters for purposes of management and stock conservation; and
- recognition of the special needs of the inshore sector in relation to investment and funding from public (i.e. EU) sources.

The adoption of such proposals, together with the earlier recommendations from the Salmon Management Task Force, would go a considerable way towards creating a more coordinated yet devolved approach to inshore fisheries management and towards restoring confidence within the inshore sector.

At the same time, inshore fisheries are likely to become subject to an increasing level of intervention through attempts to develop more holistic approaches to management of the marine environment. In Ireland a new marine policy involving an integrated approach to the development of marine resources (Marine Institute, 1996) is in its public consultation phase and a separate draft coastal zone management policy (Brady Shipman Martin, 1997) has recently been formulated. Together, these may also act as a catalyst for the development of a more coordinated approach to the management of the inshore fisheries.

7. REFERENCES

Bord Iascaigh Mhara (1999) *Irish Inshore Fisheries Sector: Review and Recommendations*, Bord Iascaigh Mhara, Dublin.

Brady Shipman Martin (1997) *Coastal Zone Management: A Draft Policy for Ireland*, Brady Shipman Martin, Dublin.

Commission of the European Communities (1992) *Regional Socio-Economic Studies in the Fisheries Sector: Ireland*, Directorate-General for Fisheries, European Commission, Brussels.

Fingleton, P. and Whelan, B.J. (1994) *The Economic Consequences of Phasing Out the Irish Driftnet Salmon Fisheries*, EIS Ltd, Dublin.

Marine Institute (1996) *Towards a Marine Policy for Ireland: Proceedings of the Consultative Process*, Marine Institute, Dublin.

Molloy, J., Condon, J. and McGarvey, P. (1996) Ireland, in *Landing statistics and biological sampling: working document for the herring assessment working group of the area south of 62° N*, Herring Assessment Working Group of the Area North of 62° N of the ICES, ICES, Copenhagen, pp. 20-28.

Salmon Management Task Force (1996) *Making a New Beginning in Salmon Management: Report to the Minister*, The Stationary Office, Dublin.

Steins, N.A. (1998) We have to keep foreigners out of our bay: top-down regulations and the strategic response of Irish fisherman, in D. Symes (ed.), *Property Rights and Regulatory Systems in Fisheries*, Blackwell Science, Oxford, pp. 216-229.

Chapter 8

FRANCE

Katia Frangoudes

1. INTRODUCTION: PROBLEMS OF DEFINITION

Defining and quantifying inshore fisheries in France is a complicated task. Various forms of classification have been developed to distinguish particular categories within the fishing fleet for administrative purposes but rarely can these be combined to create a clear, precise picture of the sector. Some forms of administrative classification are enshrined in legislation while others refer to common practice, but most are based on simple criteria such as vessel length, registered tonnage or the normal duration of the fishing trip. Other more complex classifications are to be found in academic studies dealing with the economic organisation of the fishing unit or the collective organisation of the fishermen. But unlike the distinction between artisanal and industrial forms of organisation, the separation between inshore and offshore fisheries is much more a matter of the use of space than any particular technical or socio-economic characteristics. For this reason it may be more practicable to examine the legal definitions of coastal waters in order to find a basis for identifying inshore fisheries. In summary, inshore fisheries can be defined as comprising family based, artisanal units with a limited range of operation measured in both time and space, fishing relatively close to the shore.

However, this approach does not automatically yield a simple solution. The law of 24 April 1942 (Article 8) dealing with the maritime navigation of the French fleet offers the following classification:

- small scale fisheries (*petite pêche*), normally engaged in fishing trips of less than 24 hours;

D. Symes and J. Phillipson (eds.), Inshore Fisheries Management, 139–155.

- coastal fisheries (*la pêche côtière*) involving trips of between 24 and 96 hours;
- offshore fisheries (*la pêche au large*) including trips in excess of 96 hours but excluding those engaged in
- distant water fisheries (*la grande pêche*) including vessels over 1000 GRT and vessels over 150 GRT but which are absent from port for 20 days or more.

The question here is whether to draw the line between small scale and coastal fisheries or between coastal and offshore fisheries in the search for a clear definition of inshore fisheries.

Another clue may be found in the decree of 12 February 1984, drawn up for the purpose of allocating subsidies for fleet renewal and modernisation which variously defines the size of coastal vessels as under 16 m where registered in harbours along the North Sea, Channel and Atlantic coats, under 18 m in the Mediterranean and under 12 m in the overseas territories. The variations reflect differences in fishing practice in the three geographical areas. In general, fleet statistics in France use a 5-fold classification based on vessel length (with the commonly applied descriptive terms shown in parentheses): under 12 m (small scale fisheries), 12-16 m (inshore artisanal fisheries), 16-25 m (offshore artisanal fisheries), 25-38 m (semi-industrial fisheries) and over 38 m (industrial fisheries). The term 'artisanal' indicates both the direct participation of the vessel owner usually as skipper and the remuneration of the crew through the share system. By contrast, in the semi-industrial and industrial fisheries, the vessel owner is seldom employed on board and part of the crew's remuneration is based on a system of fixed wages. The practical significance of the 12 m threshold is inherited from traditional safety regulations which applied only to vessels over 12 m.

A third and final classification, developed in 1993 by the main fishermen's organisation, *Fond d'Intervention et d'Organisation du Marché* or FIOM, recognises three distinctive groups of enterprise: first, the non-organised artisanal fisheries comprising all those units operating within the coastal zone which are not members of a producers' organisation (PO) - these constitute the majority of boats, around 6,500; secondly, the organised artisanal fisheries made up of artisanal enterprises which are members of cooperatively organised POs; and thirdly, the semi-industrial and industrial fisheries (FIOM, 1993).

Up to a point these different modes of classification are coherent. At their broadest extent inshore fisheries will be prosecuted by artisanally organised vessels under 16 m fishing within coastal waters on trips under 96 hours duration. This is, however, a very much more generous definition than that used in most other European countries. While the three typologies may serve the administrative need to produce logical sets of statistics to describe the structure of the French fishing fleet or its labour market, they are clearly arbitrary when considering the patterns of resource exploitation or systems of management. Indeed they have never been applied to patterns of production. Only since 1998 have French fisheries statistics attempted to disaggregate production data according to technical characteristics such as vessel size or fishing gears. Moreover, logbook data identifying catches by fishing area and used for purposes of quota management do not apply to small boats. As a result, the links between particular segments of the fleet, fishing grounds and type of production cannot be statistically established. This surely is a significant weakness.

Similar problems confront the definition of inshore waters when the question is approached from a purely geographical perspective. Although domestic maritime law frequently uses terms like 'coastal zone' and 'the littoral' as more or less interchangeable terms, they are not strictly defined in terms of geographical limits. In international law, the limits of territorial waters have since 1973 been generally defined as 12 nm and this has been adopted by the EU in respect of the provisional rules governing the access of non-coastal state fishing vessels. In France, Article 13 of the fisheries law no. 97-1051 of 18 November 1997, referring to the coastal zone (*bande côtière*) states that

> 'the government will produce within the next two years, and in relation to the renegotiation of the Common Fisheries Policy in 2002, a report on the specific conditions of operation within the coastal zone and particularly in reference to territorial waters, with regard to its importance for resource renewal, exploitation by locally based fleets (*flottille de proximité*), marine aquaculture and the coastal economy and employment'.

Although this law, in a sense, anticipates the outcome of the renegotiation of the access derogation and indicates a renewed interest in the economic significance of inshore fisheries, in its lack of precise definition it again draws attention to the absence of any commonly agreed definition of inshore fisheries.

Nevertheless it seems likely that the future definition of inshore fisheries will come to rely on geographical criteria and that a 'territorial approach' will play a more significant role in framing access and use rights. What is at issue in the 2002 negotiations is whether the territorial limits will achieve greater permanence and so form the basis for more robust management of the inshore fisheries - as French and British interests advocate - or, through the annulment of the current derogation, become very much weaker and subject to little more than informal local management agreements.

2. NATIONAL AND REGIONAL PROFILES

By combining the earlier classifications based on vessel size and trip pattern with the spatial limits of the coastal waters we can arrive at an operational definition of inshore fisheries and begin to develop national and regional profiles of the inshore sector. Logically, vessels under 12 m which are absent from port for less than 24 hours will normally confine their activities to within the 12 mile zone, while those spending more than 24 hours away from their home port are more likely to fish outside the 12 mile limits. Partial confirmation of this assumption is provided by the results of investigations by IFREMER which show that 80% of under 12 m boats in south Brittany operate exclusively within the 12 mile zone. By contrast, vessels in the 12-16 m range exhibit much greater diversity of fishing patterns with some operating exclusively within the 12 mile zone, some working mainly outside and others operating both within and beyond the territorial limits (Berthou *et al.*, 1997).

Since 1987, the French fishing fleet has been surveyed annually by the Interregional Maritime Affairs Directorate (DIRAM), principally to monitor the implementation of the Multi-Annual Guidance Programmes (MAGPs), and classified according to length, engine power (KW), gross registered tonnage and the average of the vessel's seasonal activities as measured on the first day of May, September and December. All vessels which have fished for at least one day in the previous two years are included, provided they have neither been wrecked, laid up for more than nine months, transferred out of fishing to other commercial activities for more than four months, nor switched to purely leisure usage. The results show a massive preponderance of small boats whose activities are likely to be limited to the inshore zone: in 1998 4,512 boats or 74% of the total active fleet were under 12 m. The broad regional breakdown of the French fishing fleet is shown in Table 8.1.

Employment in French fisheries is also reviewed annually and by the same means. Those who have been employed at sea for at least three months in the previous year are registered as regular fishermen irrespective of whether they are vessel owners or crew members. The main purpose of the registration system is to administer the maritime social insurance scheme - a direct descendant of the original insurance system established in France as early as the seventeenth century. Most registered small boat owners work their own boats either single handed or with one crew member - in contrast

Table 8.1 The French fishing fleet 1998 by vessel length (m)

	< 12 m		*12-16 m*		*16-25 m*		*25-38 m*		*38 m and over*		*Total*
	No	*%*	*No*	*%*	*No*	*%*	*No*	*%*	*No*	*%*	
1998	4,512	74	560	10	842	14	89	1	77	1	6,074
1996											
Atlantic	3,066	69	495	11	733	17	69	1	17	2	4,440
Mediterranean	1,700	87	63	3	179	9	21	1	0	0	1,963
Total	4,766	74	558	10	912	14	90	1	77	1	6,403
% change 1996-98	-5.4		+0.3		-7.7		-1.2		-7.8		-5.1

Source: Bolopion *et al.*, 2000

to crews of five or six on the offshore trawlers or up to eight men on the purse seiners. In total some 6,468 fishermen, or 37% of the total, were employed on vessels under 12 m in 1996.

Inshore fisheries are universally subject to considerable local and regional variation, but perhaps more so in France than in any of the other countries included in the present volume. French fisheries are divided between two major ecosystems - the Lusitanean Atlantic system and the Mediterranean - with their very different oceanographic conditions (bathymetry, temperature, salinity, tidal regimes and coastal geomorphology) and species composition. Local variations may reflect differences in seabed morphology and substrate and the influence of mixing of fresh and saline waters especially in proximity to major estuaries. Many of these variations are reflected in the distinctive specialisms of individual ports. To do justice to these local variations and specialisms in a brief review of France's inshore fisheries would be impossible. Attention is focussed on four macro-regional divisions - the Channel coast, Brittany, the Atlantic and Mediterranean coasts (see Figure 8.1).

There are, of course, several common denominators linking the four divisions - the diversity of fishing gears deployed in the inshore fisheries, the polyvalent nature of the inshore fleet, the preponderance of private sales rather than use of the public auction system, for example. Most inshore enterprises are based on a combination of fisheries targeting different species in different habitats with different gears at particular times of the year. In general, the degree of specialisation in design, gears and species increases with vessel length, though some small boats may be highly specialised and seasonal in their operation. The relative importance of particular gears (*métiers*) may therefore vary quite considerably at the local scale. Some gears may be banned from parts of the inshore zone usually to prevent gear

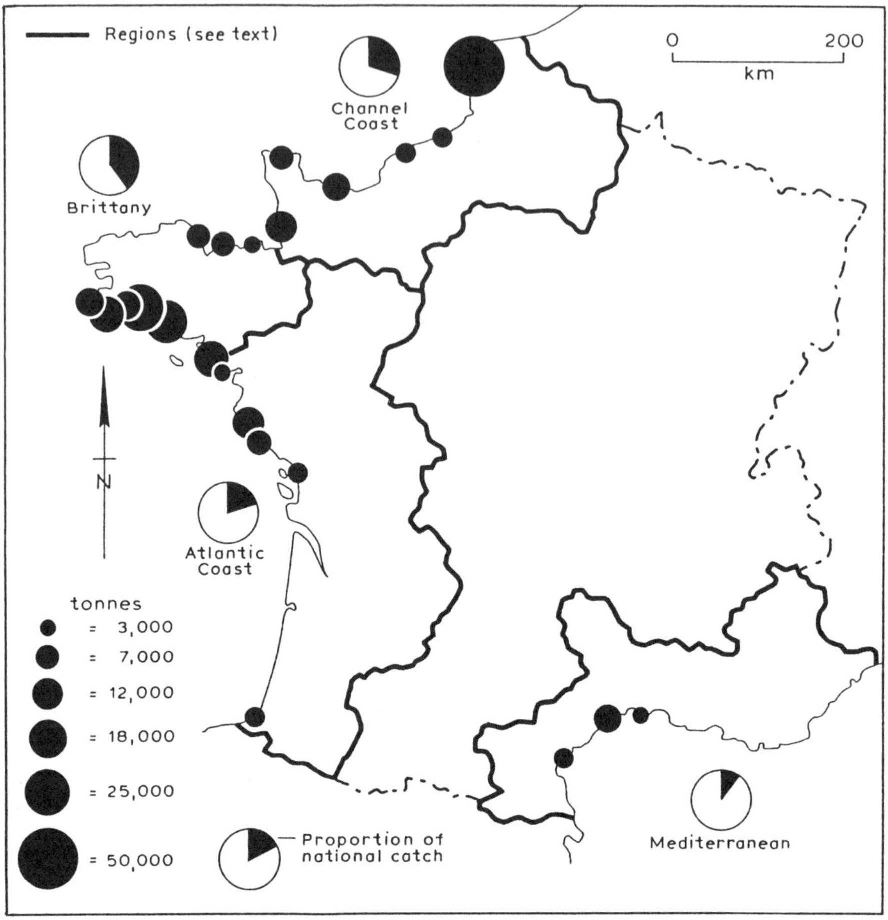

Figure 8.1 France: Regional Divisions and Principal Landing Ports, 1998 (Source: OFIMER)

conflicts: trawling, for example, is not permitted anywhere within 3 miles of the French coast.

The Channel coast, dominated by France's premier distant water port of Boulogne, provides the strongest overlap between the artisanal and industrial fisheries. Nonetheless vessels under 12 m predominate throughout the region - at least in terms of numbers - with much of their activity taking place within the 12 mile limits and often within 3 miles of the coast (Lemoine and Giret, 1991). Of all the four regions, the Channel coast not surprisingly bears the closest resemblance to the patterns of inshore fishing found in north west Europe in terms of fishing gears and species. Some variations occur within the region. In the northern part around Dunkerque the small scale fisheries combine trawling for gadoid or flatfish species beyond the 3 mile limits with the use of static gears, especially trammel nets, though there is a tendency for static gears only to be used on some of the smallest boats. Further west, on the Haute Normandie coast, though fishing practice is broadly similar, there is a stronger presence of artisanal offshore vessels and a corresponding reduction in the importance of the small boat fleet. Scallop dredging is a specialism in the Dieppe area though declining in numbers due to the overexploitation of the stocks. In Basse Normandie potting and line fishing are the two principal *métiers* with small boats of around 8 m and 45-50 KW engine power which work the inshore waters of the Cotentin peninsula.

Brittany is without doubt France's premier fishing region whether measured in terms of landings, value, employment or level of fisheries dependence. In the late 1990s it accounted for roughly half of the volume and value of all landings in France. Important distinctions are to be drawn between the north and south coasts. Along the less industrialised north coast the fishing fleet is still dominated by small, polyvalent day boats operated by one or two men (Morizur *et al*., 1992) deploying as many as four different *métier* in the same season or, indeed, on the same day. Here potting for shellfish (brown crabs, spider crabs, lobsters and crawfish) together with mussel and oyster cultivation assume a greater significance than on the Channel coast. Larger boats (*circa* 12 m) are engaged in beam or otter trawling both inside and outside the 12 mile zone, together with lining for seabass and pollock and bottom lining for conger eels.

The southern part of Brittany is the epicentre of the French fishing industry, with its two principal ports of Concarneau and Le Guilvinec ranked second and third behind Boulogne in terms of landing value; together they account for around 30% of the value of all landings in France. Despite the presence of a large fleet of artisanal trawlers (20-25 m) based in the larger ports, it is the small scale trawlers, gillnetters and polyvalent vessels fishing mainly in inshore waters which dominate the fleet structure. The most important *métiers*, combined in different forms and degrees of polyvancy, include otter trawls and small mesh nets (for bass, plaice, sole, cod, hake and red mullet), together with pots and dredges for shellfish. The inshore fleet employs more than half the region's fishermen; it tends to have lower rates of renewal with the average age of an inshore boat of 18 years compared with under ten years for the offshore fleet (Berthou *et al*., 1997).

By contrast, the stretch of the *Atlantic coast* south of Brittany as far as the Spanish border accommodates a much less developed fishing economy. The waters of the Bay of Biscay are less prolific and the strongly linear, sandy coastline especially south of the Garonne affords relatively few good natural harbours. Two thirds of the boats fishing the Bay of Biscay, both inside and outside the 12 mile limits, use otter trawls; other important inshore *métiers* include gillnetting for sole and cuttlefish, lining and potting. According to Léauté (1998: p. 357) in this region:

> 'most inshore boats operate close to their home port. Comparison of the use of space by boats from neighbouring districts shows that fishing territories overlap only to a very limited degree. Rather than reflect the need for small boats to remain close to the shore and a safe landing place, this spatial separation probably owes more to detailed knowledge of the fishing grounds transmitted over many generations'.

Several distinctive features set the fisheries of the *Mediterranean coast* apart from the other regions: first, the truncated continental shelf characteristic of the entire Mediterranean basin and rarely extending more than 40 miles off the coast; secondly, the unique 'double coastline' formed by the brackish water *étangs* separated from the true coastline by a narrow sandbar and which fringe the coast from the Spanish frontier to the mouth of the Rhone; thirdly, the wide range of commercial species dispersed in relatively small populations; and, finally, the absence of industrial fishing. To a degree, therefore, and with only a very few exceptions, France's Mediterranean fisheries are characterised by small volume landings of comparatively high priced species. Fishing activity along the mainland coast is concentrated to the west of the Rhone delta, with Sète its principal port accounting for around 40% of total landings.

In general, the Mediterranean fishing fleet suffers from a lack of capital renewal, particularly among the larger vessels (> 18 m) and especially for the fleet of light purse seiners (*lamparos*). A much higher proportion of the fleet is under 12 m (see Table 8.1) and most enterprises are organised on artisanal lines with the owner operator the most common form. The vast majority leave port each morning and return late in the afternoon with their activity largely confined within the 12 mile zone. The only major exception is the fleet of company owned purse seiners fishing for bluefin tuna beyond the limits of the continental shelf. For the inshore fleet the principal gears are: bottom trawls, targeting a mixture of species but with hake the most important; light seines involved in night fishing for anchovy, pilchard and the occasional tuna; and, closer to the shore - usually within 3 miles - and within the *étangs*, a variety of fixed gears. Here boats mainly under 8 m will fish with several kinds of gear more or less simultaneously according to season, location and target species. Not surprisingly, their daily catch will contain a variety of species but with eel, crabs, sea urchins, squid and clams of particular commercial importance. The *étangs*, which cover some 38,000 ha, are also important for aquaculture with the rope cultivation of oysters yielding 12,000 tonnes (value 2 million Euro) and mussels (8,000 tonnes valued at 1 million Euro) in 1999.

Since the late 1960s and early 1970s auctions have been established in most of the main landing ports, but private sales to merchants, restaurants, fishmongers and the public still remain important for the smaller producers.

Employment in the Mediterranean fisheries is strongly characterised by seasonality. Only 60% of fishermen work for more than six months each year and this figure may be inflated by those wishing to qualify for social welfare payments. A fifth are at sea for less than three months a year. With a continuing decline in the size of the active fishing fleet and especially in boats under 12 m which fell by 23% in the five years from 1987 to 1992, employment in the inshore sector is undergoing a quite marked contraction.

3. INSHORE FISHERIES MANAGEMENT: THE KEY PLAYERS

There are important differences in the management of inshore fisheries between the Atlantic and Mediterranean coasts of France. They reflect in part differences in the physiographic conditions and in the level of development of the Common Fisheries Policy. The absence of a fully comprehensive common policy covering the Mediterranean fisheries results essentially from the complex political geography of the region where, at present only four of the 20 coastal states are members of the EU and where there are fundamental differences in the national development policies for the fishing industry (Symes, 1999). At the same time, the truncated continental shelf has precluded the establishment of Exclusive Economic Zones and the extension of national jurisdiction beyond the 12 mile territorial limits. Apart, therefore, from this narrow coastal margin the Mediterranean is in effect an international sea. As far as Mediterranean France is concerned, the influence of the EU in fisheries management is confined largely to structural and marketing policies and to a limited range of technical conservation measures. The principal instrument of modern fisheries management - quota restrictions - is not deployed in the Mediterranean. The other major difference is the presence of an additional local player in the shape of the old established *prud'homies* whose influence is most pronounced in the management of the *étang* fisheries.

Two features characterise the organisation of fisheries management in France. The first is its strongly hierarchical structure implying a high visible administrative presence at the local level but rarely with the discretion to act independently without securing the prior agreement of the regional or national authorities. The second is the influence of syndicalism - the self-organisation of interest groups and their direct involvement in regulating professional behaviour; this, too, is structured in a clearly hierarchical form. In France, the state has tended to devolve responsibility for the detailed management of the inshore fisheries to these professionally based regional and local organisations.

To a large extent, the role of the state in the development of fisheries policy has been subsumed by the EU. As a result the state is basically responsible for translating EU regulations into French law and for establishing the minimum legal requirements for the rights to fish. Otherwise government has become concerned more with the administration of fisheries policy than with its formulation. In France, fisheries come under the administrative authority of the Department of Maritime Affairs with its broad ranging remit including coastal navigation, the policing of national and European fisheries regulations, administration of the maritime social insurance system and the preparation of statistical reports, as well as the coordination of decision making affecting different maritime activities across different departments and at different geographical scales. The integration of these multiple administrations within a specific territorial domain has made the positioning of the department somewhat unstable. But the unique corps of administration, all trained at the same military academy, provides some continuity and creates a particular mode for carrying out its responsibilities. The administration of Maritime Affairs is organised regionally and locally, with the 10 Regional Directorates (DIDAM) operating at the regional level and 20 Departmental Directorates (DDAM) at the local level. Maritime Affairs also come under the authority of the regional and departmental préfectures in order to assist the closer integration of its maritime responsibilities with other aspects of regional administration.

By far the strongest influence on inshore management is exerted by the *comités des pêches maritimes*, a professional organisation representing vessel owners, crew members, merchants and processors through a pyramidal elective structure organised at local, regional and national levels (see Fig. 8.2). It is a corporatist, para-statal organisation established in 1945, but subject to profound reform in 1992, coming under the authority of Maritime Affairs. Any participant in the fishing industry - whether in the harvesting, sales or processing sector - is a *de facto* member of the organisation, contributing to its funding through levies on the sales of fish. Up to 1992, at all levels, representation on the committees was determined by the central, regional and local administrations either directly or following nominations by the various groupings of interests involved (unions and associations). The principal reform in 1992 was the introduction of the elective principle based on union lists of vessel owners and crew members held at the local level; representatives of other groups are nominated through their own professional organisations. The parity principle of representation

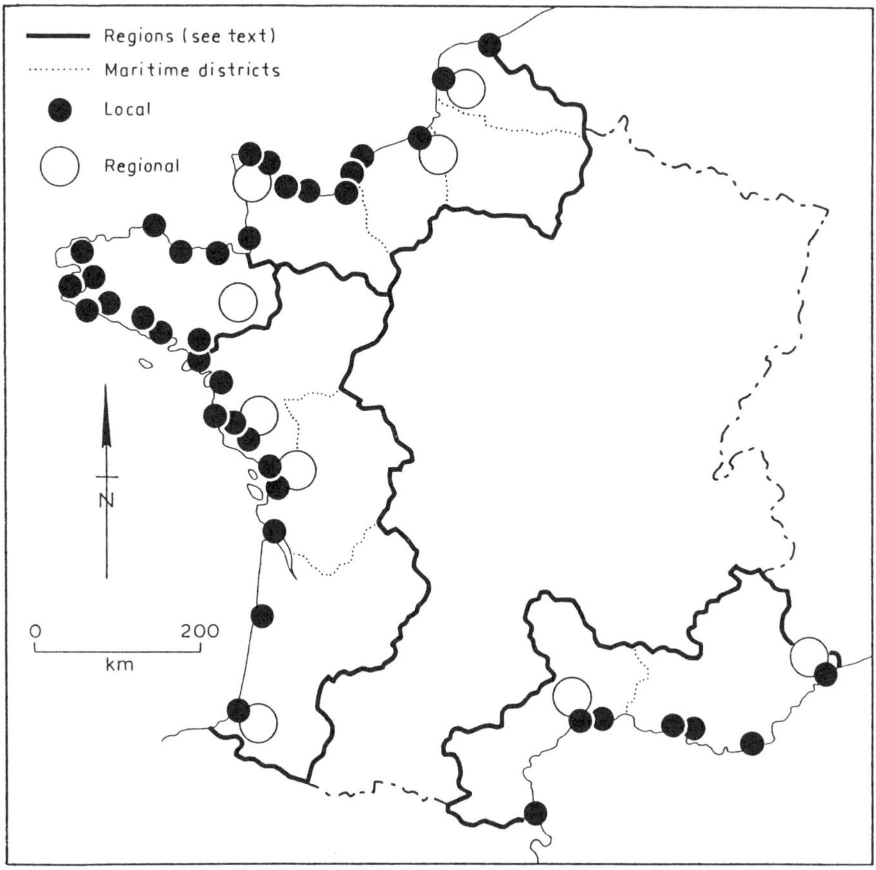

Figure 8.2 France: Comités des Pêches

for owners and workers applies throughout. The committees, elected for a period of four years, have four basic functions: representation and promotion of the general interests of the fishing industry; harmonisation of the interests of the different participating sectoral groups; improvement of the social and economic conditions within the industry; and contributions to an appropriate and well balanced system of fisheries management. At all levels, representatives of the Maritime Affairs department participate in the deliberations of the committee, but without voting rights.

Each of the 39 maritime districts in France has its own local *comités locaux des pêches* formed of three distinct constituencies: boat owners and crews; owners, managers and employees in the downstream sector (merchants, processing plants and producers' organisations); and the fishing cooperatives. Their purpose is to represent local interests and to provide advice and policy proposals in relation to local management issues. The local committees have no authority to decide on local regulations even when applied solely to local circumstances; they can only take proposals forward to the regional committees for endorsement. They are, however, responsible for implementing national and regional management measures. A further and growing function of the local committee is to facilitate meaningful dialogue with other local organisations, like the influential Chambers of Commerce which control the local auction markets but which are seen to be too commercially and insufficiently socially oriented. Raising the level of influence of the artisanal fisheries has led to increasing levels of cooperation and the establishment of joint initiatives among neighbouring local committees.

The regional *comités* occupy a pivotal position within the management system not only in relation to the overall committee structure providing an important link between local grass roots opinion and the more politically oriented national committee, but also in forging relationships between professional opinion and the levels of regional administration with powers to implement local rules for fisheries management. A key task of the regional *comité*, therefore, is to consider proposals submitted by the local *comités*, to attempt to harmonise these and, where it is satisfied, to forward any amended proposals either to DIDAM and the *Préfet de Région* for decision or to the National Committee where the proposal concerns a matter of national competence. In effect, therefore, the regional committees can exert a significant but indirect influence on fisheries management in territorial waters.

Although linked structurally to the regional and local committees, the National Committee is a very different kind of organisation. Its membership, chosen through a ballot of persons on the local lists, mainly comprises officials of the constituent unions and associations and is usually without any active professional fishermen. It is, in effect, a more overtly political organisation charged with representing the rights and interests of its national membership, though often criticised for being out of touch with grass roots opinion. In practice, the committee works closely with the Fisheries Directorate in the elaboration of national fisheries policy and also plays an active role at the European level.

Not directly linked to policy making at the national level nor to the formulation of local fisheries regulation, but increasingly influential in the overall organisation of the fishing industry in France are the 28 producers' organisations (POs) established under European legislation to advance the interests of their members in the disposition of their catches. Since 1990 POs in France have also been given responsibility for monitoring the system of quotas for the seven most sensitive species. POs are organised both regionally and sectorally with separate structures for those handling catches from offshore vessels federated through the National Association of Producers' Organisations (ANOP) on the one hand, and for POs which handle the artisanal fisheries, federated through the Federation for Artisanal Fish Producers' Organisations (FEDOPA). Altogether membership of the POs represents around 35% of the national fleet but with a much stronger proportional representation of the offshore rather than the inshore sector.

The final piece in the jigsaw of management organisations in France - but one which is confined exclusively to the Mediterranean coast - is provided by the *prud'homies*, among the oldest surviving fisheries management institutions in Europe. Similar in structure and function to the medieval guilds, they have been in existence for almost a thousand years, officially recognised in the fourteenth century, legalised by royal ordinance in 1681 and remarkably surviving the dissolution of the privileged guild system during the French Revolution. Today there are 34 *prud'homies* in France. Although their function and authority has been progressively diminished as a result of the increasing level of responsibility assumed by the state and para-statal organisations they continue to play an active role in local management. This is based on four key attributes:

- the recognition of a specific territory within which the authority of the *prud'homie* applies - to be able to fish within this territory, a boat owner must be a member of the *prud'homie* and abide by its rules; in other words, the concept of the *prud'homie* is akin to a communal marine tenure system;
- legitimacy based on a system of elected representatives (*prud'hommes*), 'men of experience and integrity' who are recognised by the community as competent to make and enforce the rules; the *prud'hommes* generally represent the different gear groups often based on the distinction between the *étang* and the open sea;
- rights of exclusion in terms of admitting new members and the principle of equity when assigning access rights to particular fisheries - in general *prud'homies* have acted to oppose the introduction of new, more efficient fishing gears which could threaten the survival of existing fishing methods;
- extended powers, especially in the field of conflict resolution, which permit the *prud'homies* to make rules which immediately have the force of community law without having to be sanctioned by other forms of jurisdiction, combined with the ability to make judgements in individual cases which are not subject to appeal.

In practice the regulatory systems developed by *prud'homies* are little more than the codification of existing uses and customs; and the *prud'homie* itself is by way of being a paternalistic, corporate institution for the management of fishermen rather than the conservation of resources, in which the guiding principle is the need to make adequate provision for all participants to derive a living from fishing. Even so, some of the common forms of regulation - deciding the times for setting the nets, mesh sizes and the suspension of fishing activities on spawning grounds - have a dual purpose of ensuring equity and at the same time conserving stocks.

Despite the fact that nominally the jurisdiction of the *prud'homies* covers both the *étangs* and the adjacent inshore waters, in reality the authority of the *prud'homie* is now largely confined to the *étangs*. Their powers to curb the use of particular gears in the inshore zone - trawling and purse seining, for example - in order to protect the interests of their own members have been abrogated by the intervention of the state. Thus there is now an internal contradiction in the principles of management between the *étang* where there is a carefully contrived, equitable allocation of resources among a clearly defined closed community, and the inshore waters where management is based on the premise of open access and where the allocation of fishing rights is not permitted either in practice or in French law. There can be little doubt that the innate conservatism of the *prud'homie* has contributed to its declining influence in competition with more modern forms of organisation which condone technological innovation but seek to limit its impacts on the resource base. Nonetheless the community based nature of the *prud'homie* earns greater respect for its rules than is customarily shown for national regulations.

4. THE INSHORE FISHERIES: SYSTEMS OF REGULATION

The regulation of inshore fisheries in France stems from four distinct sources: the EU through the provisions of the CFP, central government, the regional authorities which choose to implement the recommendations of the *comités des pêches* and, in the case of the Mediterranean, the *prud'homies*. Because of the political significance popularly attached to the derogation in respect of access to territorial waters, the extent of the EU's regulatory intervention in inshore waters is sometimes overlooked. Throughout the Atlantic region, the system of TACs and quotas defined initially in EC Regulation 1720/83 applies equally to inshore and offshore fisheries. In France, national quotas are distributed between the maritime regions and thereafter subdivided among maritime districts and the POs but not allocated to individual vessels. Nor is any part of the quota set aside for the exclusive use of the inshore fleet. As a result inshore boats fishing for quota regulated species must 'compete' with all other units for their share of the regional allocation. In practice, however, many of the fisheries prosecuted by the inshore fleet are not subject to quota restrictions: the typical polyvalent inshore boat is thus less directly affected by the vicissitudes of quota policies.

EU regulations potentially impact on France's inshore fisheries in two other ways. Since January 1995 all EU boats wishing to take part in commercial fisheries must be licensed. At present the European licensing system holds no deeper significance in terms of restrictions on movement or entry into particular fisheries though special pressure stock licences may be needed for quota regulated species. In France the need for a general licensing system had been anticipated some seven years previously with the institution of the *permis de mise en exploitation* (fishing permit) in 1988. Also European fishing vessels were required to complete logbooks indicating the location and details of the catch.

In many instances, however, the inshore fleet has been exempted from these regulations. Neither the fishing permit nor special licences are obligatory for boats under 10 m operating within territorial waters. Likewise the requirement to complete logbooks does not apply to under 10 m vessels. These exemptions are a matter of concern for the artisanal POs which handle catches from both offshore and inshore waters. According to D'Artigues *et al.* (1995) the lack of universal rules makes it difficult for POs to establish sound management practices: the POs may decide upon a management plan but they cannot enforce it.

Although largely exempt from EU regulations, the small boat inshore fleets are constrained in their activities by local rules set either by the regional administrations in response to recommendations from the *comités des pêches* or, in the case of the Mediterranean, the *prud'homie*. These local rules serve a number of purposes: regulation of the market, avoidance of gear conflicts, equality of access and resource conservation. In almost all cases it is the immediate or longer term interests of the local inshore fishermen that are being protected. In some cases the combination of local rules gives rise to a truly managed fishery as in the case of the scallop fisheries of the Bay of Saint-Brieuc in Brittany where 'every fisherman who applies to the local *comité* for a licence must also be a member of the local marketing organisation. This allows for a uniform implementation of fishing plans for the scallop beds' (D'Artigues *et al.*, 1995: p. 37). Similarly the scallop fisheries in the Bay of Brest are effectively managed by the local committee (Bailly *et al.*, 1996).

5. URBANISATION AND THE THREATS TO INSHORE FISHERIES

Of the many current issues facing inshore fisheries throughout Europe, one which has a particular resonance in France - and especially on the Mediterranean coast - is the impact of coastal tourism, marine related leisure activities and urbanisation, all of which have expanded rapidly over the last 30 years or so. Despite the introduction of quite stringent planning laws and the deliberate attempts to conserve large areas of coastal habitat through the designation of national and regional parks there is strong evidence to suggest that tourism and urbanisation are causing problems for the traditional inshore fishing industries.

Five particular problems can be identified. The first is the competition for resources between professional and 'recreational' fishermen. Although recreational fishing, including diving, is regulated, there is a widespread feeling among the professional fishermen that the rules are not being adhered to: recreational fishermen take too many fish, use illegal gears, damage the ecosystem and undermine the returns of the commercial fisheries by selling their catches informally and at low prices. Secondly, there is keen competition for space both in the harbours and at sea: any leisure activity taking place within 3 miles of the shore is likely to generate conflicts with fishing activity - windsurfing, sailing or jetskiing are all prone to disturb inshore fisheries, interfering with or damaging fixed gears.

Thirdly, urban development and the need for well drained land and the increasing demand for water is altering the delicate balance between salt and fresh water inflows into the *étangs* upon which the productivity of the fisheries depend. At the same time urban waste waters, together with industrial spoil and run-off from coastal farmlands treated with agricultural chemicals pose serious pollution risks in the *étangs*, threatening shellfish production in particular. Finally there is the conflict of imagery between the working fishing harbour and the rather sanitised, picture postcard views allegedly sought after by the visitors. Colourful fishing boats are fine, but the inconvenience, clutter and dirt associated with work on the quayside is not. Even so the late afternoon fish landings in advance of the auction have become something of a regular tourist attraction in many Mediterranean ports.

Resolving what is likely to prove a very unequal struggle between a declining traditional activity and a burgeoning tourism economy backed by strong financial interests is likely to prove difficult and will require the framework of a carefully integrated system of coastal zone management.

6. REFERENCES

Bailly, D., Boncoeur, J., Guyader, O. and Le Floc'h, P. (1996) *Economie et Droit des Ressources Naturelles Renouvelables de la Mer*, Tome 1: Etude Economique, Programme AMURE, UBO-IFREMER-ORSTOM, CEDEM, Brest.

Berthou, P., Talidec, C., Jezequel, M. and Lespagnol, F. (1997) *La Flotte de Pêche Commerciale Bretonne, Description des Métiers et des Flottilles*, IFREMER, Plouzané.

Bolopion, J., Forest, A. and Sourd, L-J. (2000) *Rapport sur l'Exercice de la Pêche dans la Zone Côtière de la France*, Ministère de l'Agriculture et de la pêche, Paris.

D'Artigues M., Catanzano, J., Lebon Le Squer, D. and Rey, H. (1995) *Les Organisations des Producteurs des Pêches Maritimes Française: Situation et Typologie*, IFREMER, RIDRV-95-20, Paris.

FIOM (Fond d'Intervention et d'Organisation du Marché) (1993) *La Filière Pêche dans la Crise*, Cellule Etudes, Paris.

Léauté J.-P. (1998) Les flottilles de pêche de l'Union Européenne dans le golfe de Gascogne vues du ciel, *Oceanologica Acta*, **21** (2), 371-381.

Lemoine M. and Giret, M. (1991) *Les Pêches Artisanales de la Manche Orientale, Flottilles et Ressources Halieutiques*, IFREMER, Plouzané.

Morizur, Y., Bertou, P., Latrouite, D and Veron, G (1992) *Les Pêches Artisanales de la Manche Occidentale, Flottilles et Ressources Halieutiques*, IFREMER, Plouzané.

Symes, D. (ed.) (1999) *Europe's Southern Waters: Management Issues and Practice*, Blackwell Science, Oxford.

PART II: COMMON THEMES: A CROSS-CULTURAL ANALYSIS OF INSHORE FISHERIES

Chapter 9

THE SOCIAL ORGANISATION AND REPRODUCTION OF INSHORE FISHING

David Symes and Katia Frangoudes

1. INTRODUCTION

We know surprisingly little about the forms of social organisation and processes of social reproduction of inshore fishing in late twentieth century Europe. Most of the evidence is fragmentary and derives mainly from anthropological studies undertaken in the remoter areas of the Atlantic Fringe in the 1960s and 1970s, focusing largely on pre-industrial forms of fishing activity. Indeed, thanks to the sustained programme of studies by the Institute of Social and Economic Research at the Memorial University in St John's, we know far more about the social organisation of Newfoundland fisheries than we do about the whole of northern and western Europe.

In the absence of recent research, it is tempting either to project the trends noted in these earlier studies in order to describe the present or to invoke parallels from the organisation and reproduction of small scale farming in Europe - an activity which may appear analogous to the artisanal forms of inshore fishing - that has been more thoroughly researched in recent years (see, for example, Gasson and Errington, 1993). Both approaches are, however, fraught with danger. Because of their cultural specificity (and the variety of fishing methods), the earlier anthropological studies contain contradictory information: it is therefore difficult to both generalise and extrapolate their findings. Invoking parallels with small scale farming assumes that the common denominators between small scale farming and inshore fishing outweigh their distinguishing features. Certainly both forms of activity are jointly characterised by a predominance of family based enterprises, high levels of risk and uncertainty, dependence upon strong seasonal rhythms and a tendency to part time engagement in a pluriactive

D. Symes and J. Phillipson (eds.), Inshore Fisheries Management, 159-175.

economy. But here the parallels end: there are some significant differences in the resource base and its means of social regulation.

A fundamental distinction occurs in the nature of property rights. In farming, property rights are specific, recognised in law, instrumental in systems of succession and inheritance and capable of renegotiation through the market. The farmer may own all the means of production outright or lease some of the fixed assets on a short or long term basis. Generally, he enjoys security of ownership or tenure. In fisheries the nature of property rights is less clearly defined. Neither the fish, nor the areas where they are to be found, are subject to normal rules of occupation and ownership. In the past the underlying assumption has been that of common property and, in theory, unlimited access to the resources of the sea. New fishing units could establish themselves at will; and fishing acted as a 'buffer economy' accommodating more participants in periods of recession and unemployment.

In reality, however, especially in inshore waters, there were many forms of 'closure' of the commons which severely constrained the notion of free access to the fishing grounds. Traditional practice sought to exert a sense of territoriality, specified access rights and 'privileged space' (Anderson and Stiles, 1973) through a variety of measures including the individual allocation of sites ('berths') for fixed gears either on a permanent or renewable basis (Anderson and Stiles, 1973), the rotation of access to the best sites (Collet, 1996) and restrictions on access to certain vessel sizes or gear types. Where there was little competition for berths, they were often treated as heritable goods. Where fishing spots were not formally allocated to individual occupation, privately transmitted knowledge of the fishing grounds worked to effect continued occupancy across the generational divide (Dufour, 1996). More recently, government intervention has further refined the concept of access rights through the introduction of licences and quotas to identify those with rights of access to particular fisheries. It has led to a disruption of established patterns of social organisation and reproduction of the inshore fisheries.

There are other distinguishing characteristics that separate small scale farming and inshore fishing as socially constructed activities - most notably the inevitable separation between home and workplace, though much less extreme than in the case of distant water fishing, and the creation of 'separate worlds' for those who fish and those who remain at home. In many instances, however, the social systems of small scale farming and inshore fishing interlock. Most inshore fishing originally developed within a context of a dual economy - most commonly farming and fishing. As a consequence, the organisation of inshore fishing was directly or indirectly conditioned by the institutional arrangements surrounding the exploitation and inheritance of the land.

2. OWNERSHIP AND REMUNERATION

Throughout most of the available literature there is widespread agreement that family and community provide the essential keys to the social organisation of inshore fishing. The principal assets - vessel, gear and, more recently, fishing rights in the form of licences and quota entitlements - are held in private ownership. Often the capital requirements of the enterprise will exceed the resources of a single proprietor or family; here ownership will be shared between two or more members of the crew, normally in an informal partnership but sometimes and perhaps increasingly in the form of a registered private company. Ownership is thus quite widely distributed within the community and shared among a fairly large number of petty capitalists. Some concentration of ownership may occur with successful families owning more than one boat and able to place their immediate kin as skippers. Only rarely will an inshore fishing enterprise be owned by 'external capital', that is by companies or individuals not directly involved in the fishing activity itself.

Replacement of lost or damaged boats and gear and, more importantly, the need to keep abreast of new technology in order to maintain one's place in the 'race to fish' has meant that the renewal of capital assets in fishing occurs more frequently than in farming. Even in the 1960s in Shetland, the lifespan of a boat was around ten years before it was sold out of the fleet as obsolete and a new and improved vessel acquired (Byron, 1980): the sale of the boat involved the dissolution of the partnership with the owners moving on to form new partnerships with the same or different partners. In some countries, notably Norway and Sweden, the constant process of renewal, involving the selling of older boats and the buying of bigger and better boats was facilitated through grants and low interest loans available from the state - a factor which helped to prevent the encroachment of external capital and company ownership within the inshore sector. As the scale of the enterprise increased and the technology became more sophisticated, there was a tendency for the greater involvement of outside capital but it is only in the distant water sector that the functions of ownership and operation were separated and company ownership of the vessel was likely to predominate.

The role of outside investors may be quite crucial in providing start-up capital to establish an enterprise but this is usually only a temporary expedient (Byron and Dilley, 1989). The primary aim of the skipper owner is to pay off such debts and to secure the financial independence of the enterprise. This may prove increasingly difficult where interest rates are high and incomes from fishing uncertain. Today, new entrants are frequently discouraged by the high initial outlays on vessel, gear and fishing rights and by the reluctance of both public and private finance to capitalise a new venture. Nonetheless, there has been a strong surge in the building of ≤ 10 m boats in recent years, marking not so much the recruitment of new enterprises to the inshore fleet as the redeployment of the capital assets of established enterprises seeking to circumvent restrictive licence and/or quota regulations applied to larger vessels. Many of these new 'inshore vessels' are capable of operating well outside the limits of inshore waters.

In common with the system of shared ownership, the means of remuneration is based on the principle of a share of the value of the catch. This is an almost universal method of remuneration in the fishing industry (Matthiesson, 1998); the only exceptions are to be found in the company owned, distant water fleets where a system of minimum wage, with or without catch related bonuses, is more common. In the inshore sector crew members - whether co-owners, kinsmen or hired hands - are ordinarily paid a portion of the catch value; the details of the system may vary but normally certain shares are set aside to cover the costs of the boat and as a return on the capital invested. In the simplest systems, the crew members receive equal shares though, with the more sophisticated fishing methods and more complex crew structures, there is a tendency for differentiation in the value of the share according to function. The share system is symbolic of partnership within the enterprise (Acheson, 1981); it provides crewmen with the incentive to fish harder and it reduces the risk to the boat owner should the returns from the fishing trip or season be poor. Equally, however, it limits the scope for profit making, returns on investment, capital accumulation and discharge of debts in the good times.

3. RECRUITMENT OF THE CREW

Inshore fisheries are normally conducted from small boats, crewed by one or two men. In general, crew sizes have fallen as capital has substituted for labour even in the small boat sector. Nonetheless, in some of the more specialised fisheries - like the *lamparo* (light seine) in the French Mediterranean - boats crewed by as many as eight men are still to be found. A combination of different fishing seasons will usually be deployed, often varying in their manpower requirements. Inshore fishing is given to periods of intense seasonal activity interspersed with periods of 'inactivity' when the owner is engaged in maintenance and repair and the crew disperses into non-fishing related employment. The basic requirements of crew selection are

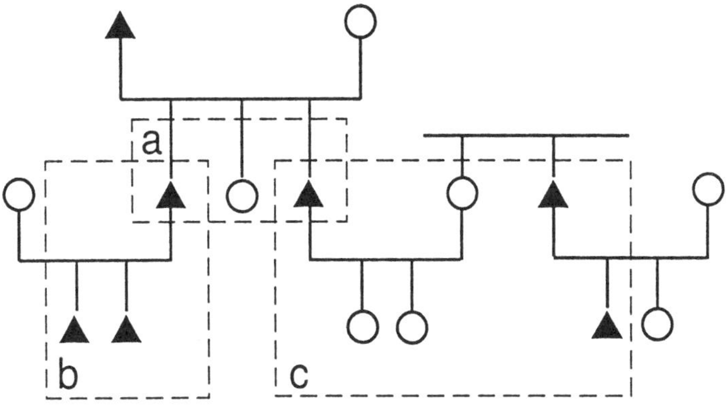

Figure 9.1 Evolution of Kin Based Crews: (a) Agnatic: All Brother Crew; (b) Agnatic: Father and Sons; (c) Affinal: Ego, Brother-in-law and Nephew

reliability, stability and efficiency (and safety) of work onboard. In most instances these requirements have to be satisfied by recruitment from within the local community.

Authors reporting on the traditional forms of inshore fishing on both sides of the Atlantic (see, for example, Anderson and Wadel, 1972; Anderson, 1979) have emphasised the role of kinship selection, though whether this arises from obligation, necessity or choice is less clear. In almost all cases, the composition of the crew is built around a regular nucleus of kin-related members. Most studies point to the primary importance of agnatic kin drawn initially from the same family household (father: son(s)) and evolving with the death or retirement of the father into brother:brother groupings (see Fig 9.1). Where the enterprise is based exclusively around close kin, the tendency will be for the crew to be renewed annually with the same members. Such a system can normally be sustained into the third generation: when a son marries and moves out of the family home to set up his own household in the locality, there is nothing to prevent him retaining his place in his father's boat. But the 'firm' will usually be dissolved when brothers fishing together have their own sons waiting to take their places in a boat - providing the means for setting up the new enterprise (essentially the capital to acquire a new boat and gear) are available.

Scarcity of agnatic kin, related to declining household size, the development cycle of the household or the exercise of occupational choice on the part of sons or brothers *inter alia*, will mean the extension of the recruitment process to include affinal kin (father:son-in-law, etc), 'friends' with no close kin relationship to the skipper, or hired men. 'Outsiders' - that is hired men with no kin or friendship ties to the skipper - will tend to be transitory members of the crew, signing on for a single season, and they will risk being displaced by the return of kin members. It is, however, rarely the case of a skipper being obliged to accept (or reject) kinsmen from particular categories, other than from within his own household. Additional crew can be actively sought out by the skipper or result from individuals activating their kinship or friendship ties in seeking a place in the boat.

Fox (1978) traces the increasing complexity of boat building and crew formation on Tory Island, north west Ireland, with the evolution from two and four men curraghs to larger sailboats requiring up to ten men to operate. Descent as a principle for the recruitment of crews was usually no more than a basis: it was never enough to build up a crew with the correct combination of skills, capital and muscle power. Beyond the core of household members, agnatic and affinal kin would be equally likely to be selected and the selection of one crew member would open up opportunities for others:

> 'If a man asked his wife's brother, then this brother would ask his uncle, and if he wanted to come in, certainly, and he would ask his daughter's husband, and they would make a crew like that until they had enough.'
>
> (A Tory Island fishermen quoted in Fox, 1978: p. 136)

The later introduction of the motor boat, requiring far less manpower, greatly reduced the need to exploit the extended family in this way.

Not all authors agree on the merits and demerits of kin based crew selection. Norr and Norr (1978) insist that skill and compatibility rather than prior relationships determine which friends or relatives will fish together, pointing out that brothers who inherit the enterprise jointly will end his partnership if there are frequent disputes over fishing activity. Byron and Dilley (1989) note in the case of Northern Ireland that skippers may actively resist any obligation to recruit kin or close friends on the simple premise that it is much harder to fire close social acquaintances. Certainly there is something of a dilemma to be faced in choosing between kin-based and more open recruitment strategies. On the one hand, the integration of capital and labour in a family firm will ensure a pooling of resources and therefore assist in the building of the family household's capital fund. Moreover, the family firm has greater resilience and 'shock absorbing capacity' (Löfgren, 1972) in adjusting to downturns in the profitability of fishing. On the other hand, a more open system, selecting crew members from a wider labour market, could help to attract those with greater aptitude and skill and so help to increase the revenues from fishing.

Throughout most of Europe, average crew size in the inshore fisheries has been falling. Several factors lie behind this trend: the impact of new vessel and gear design; the inability (or unwillingness) of the small boat owner to cover the costs of additional crew members; or the diminishing pool of willing crew members within the family or community. Improved education opportunities have opened up possibilities for occupational mobility among the young. Combined with a reluctance to become involved in a job which offers uncertain rewards for hard physical effort and which no longer competes in image or earning power with urban based blue and white collar employment, this has induced a steady outflow of young people from many fishing communities. Even among those who elect to stay in the industry, there is likely to be a more rapid turnover of crew members; the boat owner's son will tend to remain on board only long enough to serve his apprenticeship and qualify for a skipper's certificate before hoping to become his own master; though in this he may be held back by the difficulty and expense of acquiring a licence. Lack of sufficient crew may force the early retirement of the established skipper owner or at least provoke a scaling down of his activity possibly involving the sale of the boat and reinvestment in a smaller craft that can be operated single handed.

In the Mediterranean - though not yet noticeably along the Atlantic coast - the tendency has been for immigrant workers to fill the empty berths on the larger inshore boats. In the case of France, they come mostly from fishing communities along the North African coast and are employed on either a seasonal basis or on contracts of up to two years, receiving the same remuneration and social benefits as French crew members.

4. ORGANISATION ON BOARD THE BOAT

The egalitarian principle evident in the share system of ownership and remuneration commonly extends to the organisation of the crew on board the boat. For most inshore fishing activities with a settled and experienced crew there is little need for an assertion of authority: there is little division of labour and virtually no hierarchical structure. In theory, all crew members share the opportunities and risks of decision making equally. As Byron (1980) has observed all on board are deckhands. Once the decision of when and where to fish has been made by the skipper and with the tacit agreement of the crew members, work is largely a matter of well rehearsed routine with each crew member knowing his place on the boat and the tasks he must accomplish to land the catch safely. It is only where the scale of the enterprise and the level of technological sophistication have increased, where the fishing trip involves longer periods of absence from home and the complement of the crew has become larger and more stratified - as in the case of purse seining or deep sea trawling - that a more hierarchical structure of authority and clearly defined lines of command will emerge. Often in these cases, the fishing vessels are company owned and the skippers employed as 'managers' bearing full responsibility for the success of the trip.

In the inshore fisheries, by contrast, the skipper is *primus inter pares* (Norr and Norr, 1978). His status as skipper will derive from seniority, usually as the father or elder brother of the family group that forms the nucleus of the crew. His reputation and thus his ability to attract good crew members from outside the family nucleus will depend almost entirely upon his skills as a seaman, his knowledge of the grounds and his success in fishing. A successful skipper will be better able to retain a committed, experienced and stable crew.

In the past, becoming a skipper was basically a matter of seniority and experience. Today, there is an increasing emphasis on formal training, involving both 'classroom learning' and practical experience (or 'situated learning') under the guidance of an experienced skipper, in order to qualify for a master's certificate and be able to take charge of a larger vessel. Although there is a reluctance on the part of many younger fishermen to give up time at sea and return to the classroom, formal training is now widely recognised as a prerequisite for a career structure within the industry.

Much has been made in the literature of the 'skipper effect', that is the marked differences in the skills of the individual skipper as a principal cause of variations in achievement between boats fishing the same grounds (see, for example, Pálsson and Durrenberger, 1982, 1990; Thorlindsson, 1988). As Pálsson (1994) also points out, success is also profoundly influenced by the level of cohesion and experience of the crew as a whole. An experienced crew is a key to efficiency and safety on board. It is the skipper's role to select the crew and to instruct them in their tasks. He will need to balance the needs for discipline - often no more than the ability to work in harmony - and for trust and confidence, which may be earned through filial loyalty, friendship and success. Writing of Icelandic fishing in general rather than specifically on inshore fishing, Pálsson (1994) describes the relationship between skipper and crew during the fishing trip as 'authoritarian and rigid', but ashore as socialising in order to strengthen the social bond and enhance the solidarity of the crew.

While much of the skipper's reputation may rest upon personal experience and individual (or 'private') knowledge, it is common for skippers to share some of their immediate knowledge both at sea through the ship's radio and on land through informal 'clubs' of skippers meeting regularly to exchange information on a regular basis (Pálsson, 1994). Much of this shared information will tend to be generalised in contrast to the more specific knowledge concerning the 'best' fishing spots, which is likely to remain closely guarded private information. Information relating to seasonal fisheries for migrating species is more likely to be shared than that which concerns sessile and sedentary species (Acheson, 1981).

5. WOMEN'S ROLES

Within the fishing community there is commonly a strong gender based division of labour. Put very simply: men fish while women manage the household. This typical stereotyping, often explained in terms of physical skills, is culturally conditioned. Taboos frequently kept the women folk from joining the boat crew. But these had more to do with the recognition of complementary roles than with differences in stature and physique. Only in the extreme occupation (Tunstall, 1962) of distant water fishing do men and women inhabit largely separate worlds, in which husbands and wives build separate sets of ties, loyalties and activities that exclude the other marital partner. Although inshore fishing rarely involves long periods of enforced absence from home, the unsocial hours involved in some of the more intensive seasonal fisheries can disrupt normal family life. However, as many traditional fishing communities are strictly Sabbatarian, family life is normally renewed once a week.

In most traditional inshore fisheries, women had a clearly defined role to fulfil: they, in company with the older retired menfolk and young children, formed the 'shore crew' engaged in mending nets, baiting lines and preparing the catch for sale. In the case of the migratory fisheries, like the east coast herring fishery off Scotland and England, women would form mobile work groups to follow the fleet from port to port along the coast to process the catch. Where the fishing enterprise formed part of a dual economy of farming and fishing, the wife was commonly left to run the farm during the seasons of intensive fishing activity when the men were away fishing. Her work roles would often involve three competing areas of activity: home and family, the farm and the fishing enterprise.

With the development of factory based processing of the catch, the complementarity of the boat and shore crew diminished. Women moved into the processing plants becoming part of a seasonal, wage earning proletariat and providing a valuable source of additional income for the household. Despite this separation between family enterprise and factory employment, the skippers' wives were still expected to undertake important, if invisible, and largely unrewarded, tasks on behalf of the fishing enterprise - keeping accounts, dealing with banks, insurance and tax returns, ordering spare parts etc, while the husbands were at sea. Such activities have become so central to the success of the fishing enterprise that in France there is a growing demand for short training courses to develop the basic skills involved in running a small family business. Thus, women have become increasingly implicated in the business side of the enterprise. Not surprisingly, therefore, the old Scottish saying that 'no man can be a fisher and lack a wife' retains something of its currency today. In several countries, but most notably in Norway and France, women have also become involved in political action to secure better conditions of employment in the fishing industry both for boat owners and crew members.

The traditional stereotype of the woman's role within the fishing community is breaking down. According to Pettersen (1996), in Norway there is a growing tendency - though not yet a very pronounced one - for women to become directly involved in fishing as part of the boat crew in family run enterprises. The same has also been noted in France, especially in the small boat sector where the skipper-owner may be facing difficulties in recruitment and in meeting the wage and social welfare payments for hired crew. At the same time, there is increasing occupational separation between men and women reflecting a growth in occupational opportunities for women outside the family enterprise and beyond the fishing sector itself. Through their employment, women may well be creating important links with other branches of the local economy - most notably perhaps, in the development of marine based tourism. And, in common with other forms of family firm - as in the case of family farming for example - fishermen's wives are being brought into the firm as partners, either to reflect the importance of their roles in managing the shore based side of the business or for fiscal reasons to reduce the tax burden on the firm (Hoefnagel, 1996).

6. SUCCESSION AND INHERITANCE

Succession and inheritance of the inshore fishing enterprise differ in a number of respects from certain other family businesses. In farming, for example, succession and inheritance may be delayed until the death of the male spouse - and even then the farm may continue in the name of his widow. Moreover, a common strategy for protecting the viability of the farm business involves the succession of only one heir, usually the eldest son but not infrequently the youngest (Symes, 1972). By contrast, succession to management of the fishing enterprise can occur quite early in the household cycle as the skipper (father) succombs to the physically demanding and potentially stressful work at sea and retires to take on shore based employment, leaving the eldest son to assume responsibility for fishing.

Furthermore, the fishing enterprise does not conform to the concept of hereditable property in the same way as real estate. It is subject to regular renewal of the principal asset (the boat); nor is it readily divisible among competing heirs. Intergenerational transfers will therefore customarily imply the joint inheritance of the enterprise, involving all surviving sons, with the eldest assuming the role of skipper. Where ownership involves a partnership extending beyond the immediate kin group, the succeeding son(s) will inherit only their father's share. Division of the inheritance is most likely to occur when a third generation of sons is ready to take their place within the boat.

Modern forms of ownership, where for example outside capital is involved or where wives have been incorporated within the firm, will tend to complicate the transfer of the enterprise. Company ownership may provide the basis for continuity of the firm which would otherwise be disrupted by death or divorce. Problems can arise when the widow (or divorced wife) seeks a settlement based on her portion of the joint property including the assets tied up in the boat (Hoefnagel, 1996). Similarly, the burden of inheritance and/or capital gains tax may force the liquidation of viable firms or incur heavy debt obligations in cases where money has to be borrowed to pay off tax liabilities. Such problems are exacerbated in cases where, for tax purposes, the value of a fishing licence and/or individual quota entitlement is included in the value of the vessel.

7. THE FISHING COMMUNITY

Traditional patterns of social organisation were largely developed within the context of a local community formed of a nexus of economic, social and cultural relations which provided a continual source of capital, labour and social security through which the system could be reproduced. As the preceding sections will have made clear, this local community was an integrated social system built very largely on agnatic and affinal kinship ties and interactions with close neighbours, which helped to generate enduring, egalitarian and essentially reciprocal economic and social relationships. It was also a closed occupational community in which most, if not all, of its members shared a close affinity with fishing.

In common with the fate of the traditional rural community, this picture of the fishing village has been greatly transformed. In some cases it has been replaced by the image of a 'company town' shaped by the incursion of a large scale, capital intensive processing industry, where economic linkages are internalised within a corporate organisational structure - through the vertical integration of processing plant and fishing fleet - and where hierarchical social structures have replaced the egalitarian relationships of the traditional fishing village (Jentoft, 1996).

This is an extreme situation, though one which is recognisable in several coastal states around the North Atlantic basin. More commonly, the social and spatial boundaries of the fishing community have been blurred by demographic processes and the replacement of indigenous fishing families by commuters, retired persons and by holiday homes (Nuttall and Burnett, 1998). Fishing populations have thus become encapsulated within more open but socially segregated coastal villages. Moreover, the former tightly clustered, localised kin based networks of fishing crews have been both diluted by the introduction of well qualified 'outsiders' and dispersed by greater individual mobility. Indeed, in cases where the geographical base of the fishing operation has been moved to a larger, more accessible harbour, the fisherman has himself become a commuter.

As the fishing communities become socially more diverse, so too marriage patterns of fishermen become both socially and spatially diffuse. Instead of marriages between fishing families in the same or neighbouring communities, today incoming wives originate from a wide variety of social backgrounds, educational attainment and occupational skills. As a consequence, the socialisation of children is less strongly oriented towards fishing, reinforcing the experience of post-primary education - which focuses on the outside world and commonly takes place beyond the boundaries of the local community - and repeating the experience of upland farming communities where intergenerational continuity has weakened (Symes and Appleton, 1986).

The territorial notion of the fishing community has thus been replaced by a functional community of well defined groups who share a common interest in a particular fishery but who have no other specific geographical or social ties (Jentoft *et al.*, 1998). Increasingly their patterns of interaction are governed by contractual relations rather than by reciprocal kinship obligations. At the same time, the notion of local systems of self-management of the inshore fisheries have been largely superseded by state-led, bureaucratic institutions.

8. SURVIVAL STRATEGIES

Even at the best of times, inshore fishing is likely to prove a precarious livelihood. Pre-industrial fishing communities had learned to cope with the endemic uncertainties of the marine environment, the weather and the behaviour of the fish. They were commonly enmeshed in a largely self-sufficient dual economy, in which small scale farming or garden cultivation provided a means of subsistence whilst fishing offered both a dietary supplement and a source of cash income. Not only was knowledge of the local ecosystem used to develop complementary fishing seasons, but the emphasis could be switched between farming and fishing as circumstances dictated. Careful husbanding of scarce capital resources could enable modest investments in new technologies and new skills could be acquired to allow diversification of the family enterprise.

In more recent times, risk and uncertainty have blossomed into crisis, compounded by new forms of pressure arising from the depletion of resources, the introduction of new systems of regulation to reduce fishing effort on vulnerable stocks, the privatisation of property rights and the globalisation of food markets *inter alia*. At the same time, the customary coping strategies have been breaking down. The dual economy of farming and fishing has been largely abandoned with the separation of the two activities into largely discrete full time specialisations. With the increasing scale of enterprise and expensive technology it was no longer economic to fish only for a short season and then lay up the boat for the rest of the year. The younger generation was less willing to accept the limited horizons and endure the submerged identity of part time employment (Symes, 1982). Such changing attitudes were accompanied by a weakening in the traditional processes of social reproduction of inshore fishing: kinship bonds became attenuated through a declining family size, the dispersal of the household through migration, the growing influence of formal education opening up new vistas of occupational choice and the increasing social differentiation within the community. Problems of survival of the family enterprise are not always linked to scarcity of resources and poor returns from fishing. In Mediterranean France, for example, where the inshore fisheries are still capable of providing a good living, even by modern standards, the issue has more to do with changing intergenerational attitudes to the image of fishing as a professional activity and therefore with problems of recruitment of crew members from within the family or community.

To a degree, modern survival strategies are adaptations of the traditional means of combating risk and uncertainty. But few studies have confronted this issue. Pettersen (1996) has highlighted four alternative strategies, each involving the redeployment of women's roles, in her study of Lofoten households during the northern cod crisis:

(i) *expansion*, where the attempt was made to increase fishing related activity through the expansion of the family enterprise, involving an extension of the wife's involvement in the family firm at the expense of non-family crew members; the husband's activity, meanwhile, was increased either through a further combination of fishing seasons, the deployment of a second boat or paid work crewing on a larger fishing vessel;

(ii) *diversification*, involving an increase in paid employment outside fishing, principally for the wives, which acted as a buffer against fluctuations in the returns from fishing, while the boat owner sought to find a more cost effective use of the boat, including a possible reduction in operations;

(iii) *retrenchment*, or survival on the basis of stable or reduced fishing activity, often combined with social security payments;

(iv) *withdrawal* from fishing activity altogether, selling the boat and relying wholly on unearned income (pensions etc); this latter strategy was confined mainly to households on the threshold of retirement but it also included a number of younger households with little formal education and limited work experience.

Another source of challenge to conventional inshore fisheries comes from coastal aquaculture, including competition for space, labour and markets and risks of pollution to the coastal environment and disease transmission from caged to wildfish stocks. Aquaculture implies significantly different patterns of ownership, social organisation and regulation (van der Schans, 1996). Modern fish farming is characterised by industrial forms of ownership, heavily supported by external sources of capital and controlled principally by terrestrial planning systems. Its development has been actively encouraged by government policy (Steins, 1998) in some cases to the detriment of established inshore interests.

The state's role in sustaining the inshore fishing sector has, in fact, been ambivalent, torn between support for structural rationalisation and the need to protect employment in remoter coastal districts. Ostensibly, the continued designation of 6 mile limits for the exclusive use of coastal state vessels is intended to afford protection to local inshore fishing interests. Similarly the exemption of small boats, usually under 10 m, from certain obligations relating to licensing, quota allocations and the registration of landings would appear to be intended to support the inshore sector though, in truth, it may have more to do with ease of administration. By contrast, the introduction of rules governing minimum levels of participation in the fisheries as a qualification for licences, quotas and grant aid (Eikeland, 1998) acts to disadvantage the part time, usually inshore, fisherman. Moreover, the introduction of individual transferable quotas has, in some cases, precipitated a decline in the number of independent small boat enterprises where owners have taken advantage of windfall profits and sold their quota entitlements to larger vessel owners.

9. CONCLUSION

Inshore fishing is one of the oldest economic activities. In the past it was almost exclusively a small boat enterprise relying upon low levels of technology and socially reproduced through locally defined kinship ties. Today the established systems of social organisation and reproduction are threatened internally by widespread changes affecting the structure of the domestic household and the dilution and dispersal of kinship ties, and externally by resource depletion and the usurpation of local management responsibilities by state-led management institutions. Although inshore fishing has been weakened by such developments, it remains a numerically significant sector of Europe's fishing industry - even if increasingly marginalised in terms of its contribution to total production. Professionalisation of the inshore sector has also made its mark in recent years with the insistence on formal training and considerable sea time in order to qualify for a skipper's certificate and the growing separation of full time 'professional' fishermen from sparetime, recreational fishermen in terms of access to resources through licensing and quota schemes (see Chapters 2-3). These latter developments also make it hard to gain entry to the industry from outside, except as a crew member, simply because of the scarcity and high cost of fishing licences. Inshore fishing is increasingly a closed occupation. Family based enterprises prevail, still reliant upon a nucleus of close kin but ever more in need of supplementing domestic sources of capital and labour from outside the nuclear household.

10. REFERENCES

Acheson, J.M. (1981) Anthropology of fishing, *Annual Review of Anthropology*, **10**, 275-316.

Anderson, R. (1979) *North Atlantic Maritime Cultures*, Mouton, The Hague.

Anderson, R. and Stiles, G. (1973) Resource management and spatial competition in Newfoundland fishing: an explanatory essay, in P.H. Fricke (ed.), *Seafarer and Community: Towards a Social Understanding of Seafaring*, Croom Helm, London, pp. 44-66.

Anderson, R. and Wadel, C. (eds.) (1972) *North Atlantic Fishermen: Anthropological Essays in Modern Fishing*, Memorial University of Newfoundland, Institute of Social and Economic Research, St John's.

Byron, R.F. (1980) Skippers and strategies: leadership and innovation in Shetland fishing crews, *Human Organization*, **39** (3), 227-232.

Byron, R. and Dilley, R. (1989) Social and micro-economic processes in the fishing industry, in R. Jenkins (ed.), *Northern Ireland: Studies in Social and Economic Life*, Avebury, Aldershot, pp. 56-67.

Collet, S. (1996) Leviathan management or customary administration: the search for new institutional arrangements, in K. Crean and D. Symes (eds.), *Fisheries Management in Crisis*, Blackwell Science, Oxford, pp. 99-109.

Dufour, A-H. (1996) Management and practice in the small scale inshore fisheries of the French Mediterranean, in K. Crean and D. Symes (eds.), *Fisheries Management in Crisis*, Blackwell Science, Oxford, pp. 110-115.

Eikeland, S. (1998) Flexibility in northern waters, in D. Symes (ed.), *Northern Waters: Management Issues and Practice*, Blackwell Science, Oxford, pp. 90-102.

Fox, R. (1978) *The Tory Islanders: A People of the Celtic Fringe*, Cambridge University Press, Cambridge.

Gasson, R. and Errington, A. (1993) *The Farm Family Business*, CAB International, Wallingford.

Hoefnagel, E. (1996) Trade in fishing rights in the Netherlands: a marine environment market, in K. Crean and D. Symes (eds.), *Fisheries Management in Crisis*, Blackwell Science, Oxford, pp. 61-70.

Jentoft, S. (1996) The conversion of fisheries dependent regions in Norway, in *The Conversion of Fisheries Dependent Areas: Aims, Experience, Prospects*, Report of round table meeting, 17th May 1995, European Commission, Brussels and Luxembourg, pp. 60-69.

Jentoft, S., McCay, B. and Wilson, D.C. (1998) Social theory and fisheries co-management, *Marine Policy*, **22** (4), 423-436.

Löfgren, O. (1972) Resource management and family firms: Swedish west coast fishermen, in R. Anderson and C. Wadel (eds.), *North Atlantic Fishermen: Anthropological Essays in Modern Fishing*, Memorial University of Newfoundland, Institute of Social and Economic Research, St John's, pp. 82-103.

Matthiesson, T. (1998) Sharing revenue in Icelandic fisheries, in D. Symes (ed.), *Northern Waters: Management Issues and Practice*, Blackwell Science, Oxford, pp. 130-141.

Norr, J.L. and Norr, K.L. (1978) Work organization in modern fishing, *Human Organization*, **37** (2), 163-171.

Nuttall, M. and Burnett, K. (1998) The negotiation and management of crisis in the Scottish fishing industry, in D. Symes (ed.), *Northern Waters: Management Issues and Practice*, Blackwell Science, Oxford, pp. 238-246.

Pálsson, G (1994) Enskilment at sea, *Man*, **29**, 901-927.

Pálsson, G. and Durrenberger, E.P. (1982) To dream of fish: the causes of Icelandic skippers' fishing success, *Journal of Anthropological Research*, **38**, 227-242.

Pálsson, G. and Durrenberger, E.P. (1990) Systems of production and social discourse: the skipper effect revisited, *American Anthropology*, **92**, 130-141.

Pettersen, L.T. (1996) Crisis management and household strategies in Lofoten: a question of sustainable development, *Sociologia Ruralis*, **36** (2), 236-248.

Steins, N.A. (1998) We have to keep the foreigners out of our bay: top-down regulations and the strategic response of Irish fishermen, in D. Symes (ed.), *Property Rights and Regulatory Systems in Fisheries*, Blackwell Science, Oxford, pp. 216-229.

Symes, D.G. (1972) Farm household and farm performance: a study of twentieth century Ballyferriter, southwest Ireland, *Ethnology*, **11**, 25-38.

Symes, D.G. (1982) Part time farming in Norway, *GeoJournal*, **6** (4), 351-354.

Symes, D. and Appleton, J. (1986) Family goals and survival strategies: the role of kinship in an English upland farming community, *Sociologia Ruralis*, **26** (3-4), 345-363.

Thorlindsson, T. (1988) The skipper effect in the Icelandic herring fishery, *Human Organization*, **47**, 199-212.

Tunstall, J. (1962) *The Fishermen*, McGibbon and Kee, London.
van der Schans, J.W. (1996) Colonizing the coastal frontier: governing marine salmon farming in Scotland, in C. Bailey, S. Jentoft and P. Sinclair (eds.), *Aquacultural Development: Social Dimensions of an Emerging Industry*, West View Press, Boulder, pp. 143-168.

Chapter 10

INSHORE FISHERMEN: CULTURAL DIMENSIONS OF A MARITIME OCCUPATION

Rob van Ginkel

1. INTRODUCTION

European fisher folk often constitute close-knit occupational communities. Fishing is not merely a job, it is also a way of life. Fishers consider themselves to be engaged in the same kind of work which they view positively. On the basis of their 'work worlds', fishers at local level fraternise and identify with one another. The cultural forces that facilitate such group identity include: esoteric knowledge, skills and expertise; extreme or unusual work demands; consciousness of kind; pervasiveness whereby social relationships meld the realms of work and leisure; ideologies or norms, values and perspectives that apply to, but extend beyond, work routines and that bestow positive self-images and social value on the tasks; standards for proper and improper behaviour; work codes surrounding relatively routine practices; mental maps giving primacy to the occupational group as a reference group; consistent cultural forms - including a special language or argot, shared rituals and taboos, occupationally unique symbols, myths, stories and jokes; and compelling accounts attesting to the logic and value of these cultural forms (van Maanen and Barley, 1984; Trice, 1993).

Fishers derive considerable satisfaction from their work and they are extremely proud of their identity as fishers, even when they fish only part time. As Taylor (1977, 1981) shows with respect to an Irish community, fishing can be of modest economic importance yet be the subject of considerable cultural attention, and provide a source of communal and personal identity. In the Swedish fishing village of Öckerö there is a common awareness of sharing an unusual way of life, a strong sense of community, intense loyalties to boats with accompanying status rivalries (Byron, 1993). Danish west coast fishers

D. Symes and J. Phillipson (eds.), Inshore Fisheries Management, 177–193.

assign a high priority to their occupation as a means of livelihood (Rasmussen, 1974). On the Shetland island of Whalsay, fishing is perceived 'as the very quintessence of local identity' (Cohen, 1987). With respect to Galician fishers, LiPuma (1992: p. 50) remarks that

> 'being a fisherman is not a job (i.e. labor is not fungible) but an occupation and mode of labor which is inseparably tied to a person's identity. A fisherman's self-image, his presentation of self in the public arena, the resources that he uses to define his place in the community, his kin relations and village friends, and the trajectory of his family are intrinsically bound to fishing. It is this network of relations and meanings, determined by life in the fisheries, which defines the ground in terms of which people construct their subjectivity'.

In strikingly different settings, one may encounter among fishermen remarkably similar ideas concerning work ethos, egalitarian ideology, rhetorics and concepts of independence, self-reliance, freedom and so forth. McGoodwin (1990) perceives these attitudes as psychological coping mechanisms in an occupation where physical danger and economic uncertainty are rife. Owing to the fact that fishers adapt to similar environments and face corresponding problems, fishing communities and fishing cultures the world over would seem to have a number of characteristics in common. Such commonalities are often explained in functionalist or ecological terms.

However, on closer inspection, a good deal of difference exists between fishing systems and the ways in which they are embedded in encompassing socio-economic structures and cultures (Pálsson, 1995). Löfgren (1977: p. 235) deems functionalist or ecological explanations 'too simplistic': 'The dependance upon a marine common property resource no doubt influences the fisherman's belief and behaviour, but how this dependance is structured into cultural forms depends upon the economic and social structure within which the fisherman is operating'. Thus, fishing may be for subsistence, for the market or a combination; it may be subject to quota regulations, licensing or other measures; it may be small scale, medium scale or large scale; it may be inshore, mid-water or offshore; it may be seasonal or year round; it may be full time or part time; it may be owner operated, done by crewmen who are hired by land based shipowners, or vertically integrated. Different modes of production entail different social relations, rationales and motivations. But even spatially and socio-economically proximate fishing communities can exhibit vast cultural differences (van Ginkel, 1993). Factors like boat size, ownership structure, degree of indebtedness, number of crew members, variation in fish species, species pursued (e.g. migratory or sedentary), technology and gear used create differences in mental maps, cultural rules, practices, styles, goals and aspirations. Similarly, there is diversity with respect to preferred modes of regulation, institutional arrangements, organisations, management regimes and policing. These differences are closely linked to occupational and fleet structure, technology, division of labour, social practices and cultural backgrounds (E. Vestergaard, 1996).

Though this cultural variety in fishing societies is acknowledged and appreciated, this chapter focuses on a number of family resemblances found in the occupational cultures of European inshore fishers and fishing communities. The data are predominantly gleaned from maritime anthropological literature. Its aim is to point to some common themes and patterns in the current worldviews, lifestyles and concerns of European inshore fisher folk, and the difficulties they face. It deals not with long term developments but focuses on contemporary issues and features.

2. BECOMING A FISHER

Each prospective fisher must learn the cultural and behavioural modes of the occupational community of fishers to which one is a newcomer (either as a child or as an adult). Through enculturation or socialisation, an outsider 'learns the ropes' of fishing and becomes an insider. This process is not limited to the mere performing of tasks; it includes internalising the norms, values, attitudes, interests, knowledge and skills necessary to become an accepted member of the occupational group, to do the job properly, and to legitimise the work world. Compatibility with the crew's ideology is an important factor.

Continuity in family firms is highly valued. When an adolescent fisherman's son shows an interest in becoming a fisher, he will occasionally join his father on a trip to sea, simply being aboard and lending a hand if and when possible. This way, the father gets to see whether his son has 'the hang of it'. But prior to his initiation as a fisher, he has often already been prepared for the job on land. For example, in Burra (Shetland Islands), '[t]he lore and skills of fishing and seamanship are learnt in play, and the introduction to a boy's first job as apprentice is long and gradual' (Byron, 1978: p. 3). The fisher-to-be has gained a 'reputation' even before joining a crew.

As a rule of thumb, the process of becoming a fisher involves informal and gradual enskilment (sometimes in addition to formal education). Fishers get an 'on the job training', and have to learn as apprentices with the entire crew as a collective tutor. After the initial rites of passage, young crew members are usually quickly accepted and receive a full share. They have to show flexibility, compatibility, a willingness to work hard and absolute loyalty to the boat (Wadel, 1972; Baks and Postel-Coster, 1977; Byron, 1994). In many maritime communities 'women exert as an important influence on the shaping of a fisherman as their husbands' (Löfgren, 1984: p. 293). Fishermen's wives have an important part in giving their sons (and occasionally daughters) an outlook on life which makes fishing as an occupation attractive to them (Nadel-Klein and Davis, 1988). But, of course, there is much variation across generations and across and within cultures in maritime socialisation. In many communities, fishing has become a less attractive occupation.

3. EGALITARIANISM AND LEADERSHIP

Occupational communities of fishers are often characterised by an egalitarian ideology (Cohen, 1987). As Peace (1991: p. 5) remarks concerning an Irish inshore fishing community: 'The idea that every man is as good as his neighbour is a consistent thread informing most important codes of interpersonal conduct'. Egalitarianism is a moral principle. This does not mean that actual social relations are necessarily egalitarian. Nonetheless, a good deal of camaraderie is reported concerning life aboard fishing vessels. The ethic of cooperation emphasises equivalence and plays down status differences between crew members (Byron, 1978, 1986). Usually, the skipper is a *primus inter pares* (Löfgren, 1972). The share system of remuneration - in which all crew members are co-adventurers and share risks - increases the commitment and labour motivation of the crew, promotes cooperative behaviour and fosters a spirit of egalitarianism (Acheson, 1981; McGoodwin, 1990). The fact that crew membership and kinship, friendship and community ties are often intertwined also helps to mitigate status differences and maintain an egalitarian ethos. Such multiple ties contribute to cohesive relationships marked by unity, solidarity and mutual regard (LiPuma, 1992). Moreover, the seniority of a father or elder brother makes for a 'natural' hierarchy that facilitates playing down lines of command.

With egalitarianism being the norm, skippers have to solve the dilemma of direction and leadership versus teamwork and voluntary cooperation, especially so since leadership is often met with suspicion (Byron, 1975, 1986). They have formal power but rely on informal authority based on personal relationships and legitimated by their reputation for skill and success (Byron, 1980). Skippers rarely give direct orders or shout. A nod, a cue or a hint often suffices, especially in experienced crews. Instead of a command structure, there is an 'ideology of wordless cooperation' (Byron, 1986: p. 98).

But impression management or ideology is not practice. Even if intracrew relationships would appear to be egalitarian, there are status and prestige differences. For example, a cook usually enjoys lower prestige than a skipper. This is often related to age differences; as a rule of thumb, cooks are the youngest crew members (Prattis, 1973; Knipe, 1984). A skipper or a skipper-owner has certain privileges and he may receive a higher share than 'ordinary' crew members. He does not have to participate in the most disagreeable tasks. In inshore fishing, however, and especially aboard small vessels with a small crew, the degree of specialisation is low, the division of labour simple, tasks are shared or exchanged through rotation or alternation and an egalitarian ideology is more strongly adhered to than aboard larger boats.

4. COOPERATING INDIVIDUALISTS

In many fishing communities, an ethos, rhetoric and idiom of freedom and independence prevails (E. Vestergaard, 1996). Independence and self-employment are highly valued, particularly among small scale, inshore fishermen who own their own boats and work alone or with a small crew of kin. There would seem to be a paradox in many descriptions of maritime communities: 'On the one hand the fisherman is portrayed as a secretive individualist and gambler, on the other hand the social ethos of the fishing community is characterized as egalitarian and cooperative' (Löfgren, 1977: p. 234). Despite the individualism ascribed to them, there is overriding evidence that fishers do cooperate. In many cases fishers are 'cooperating competitors' (van Ginkel, 1996). As Hanna and Jentoft (1996: pp. 47-48) explain, 'When resource users compete, their interaction is contained within rules of the larger society. Users are guided by the ethical principles, social duties, and responsibilities that prevail within the community or ethnic group to which they belong. Thus, competition may evolve without causing social disruption and disorder. In fact, competition and cooperation should not be regarded as mutually exclusive activities, because competition cannot take place without cooperation'. At local level, face-to-face relationships sometimes make for a 'moral economy' in which transactions between fishers and merchants or processors are to a certain degree mitigated by shared community norms and values.

The interplay of competition and cooperation also holds true in access to and use of resources. Fishers have in many cases been cooperative where it concerns 'community-based, spontaneously developed and informally organized' regulatory measures to restrict fishing (Jentoft and Kristoffersen, 1989: p. 355). This capacity of self-regulation should be acknowledged and used in fisheries management. It would avoid contestation and evasion of rules and enhance legitimacy of and compliance to regulations. Social control also plays a role in observing rules and regulations (Frangoudes, 1996). However, it would be wrong to assume that all fishers are 'noble commoners' and that their prime motivation is conservation of marine resources. But it is equally wrong to view their behaviour as innately selfish and rapacious.

When facing problems in the exploitation of marine resources or when denied access to them, fishers often resort to collective action to counter the situation. Scottish fishers had a reputation of being fiercely independent, but Knipe (1984: p. 98), notices 'a growing spirit of collectivism'. Indeed, Scottish, English, Dutch, French and Spanish inshore fishers have staged several protests, boycotts and blockades in recent times (LiPuma, 1992; Delbos and Prémel, 1996). However, though fishers can be cooperative, formalising cooperation into voluntary associations and other organisations is not always successful. This is often due to the fact that locally, regionally or nationally, fishers are a heterogenous group expressive of contrasting and often conflicting ideologies and behaviours (Jentoft and Davis, 1993; van Ginkel, 1996). They also have little withholding power and a weak bargaining position *vis-à-vis* traders and politicians. On the other hand, time-honoured institutions like the *cofradías* in Spain (Alegret, 1996) and the *prud'homies* in the French Mediterranean (see Chapter 8) have played an important role in regulating fisheries and resolving conflicts for many years. Meltzoff and LiPuma (1986: p. 695) argue that Spanish littoral fishers, following adverse consequences of the Law of the Sea, have begun to establish or join associations 'that integrate familialism with an international outlook. The associations have two tiers: one that preserves and encourages individual ownership and another that defends the interests of fishermen in national and international forums'.

5. COMPETITION AND INFORMATION MANAGEMENT

Being competitors, fishers have adopted specific information and communication management stratagems. They often attempt to monopolise a particularly good fishing spot through tactics of secrecy and deception: 'Such claims, which fishermen are continually trying to make, are defined by relations of competition: the establishment, maintenance and adequacy of exploitative strategies in the hope of establishing economic advantages over rival crews' (Byron, 1975: pp. 152-153). They exchange this information with kin, close friends and possibly fellow villagers only, while being reticent or even deceptive *vis-à-vis* 'outsiders' (Löfgren, 1972; Jorion, 1978; Moustgaard, 1984; Byron, 1986; Cole, 1991; Pálsson, 1991; Peace, 1996; Mondardini Morelli, 1998). With respect to a Portuguese fishing village, Johnson (1979: p. 246) explains: 'Given the intensive competition between boats and the efforts to guard one's own information while discovering the information of others, a social climate rife with secretiveness, lying, avoidance, and general suspicion is generated'. In the course of his career a skipper accumulates a wealth of knowledge concerning specific fishing areas. In a process of trial and error, he develops a cognitive map of good fishing locations, wrecks, reefs, currents, depths, breeding habits and migration patterns of fish, the behaviour of other fishers and so on. This kind of knowledge is highly valued, as is evident in the way fishers treasure their mental maps and private log books. A skipper will most likely not share (all of) this information with others but will pass on this cognitive capital to his successor(s) (Jorion, 1978; Acheson, 1981; Peace, 1991; Dufour, 1996; Alegret, 1998).

However, McGoodwin (1990: p. 17) points out that there is a 'skipper's paradox': skippers must maximise their gains of valuable information and minimise their loss of it. At the same time, they must be 'socially participant cooperators and competitors'. In other words, they have to balance the dilemma of being competitive and secretive on the one hand, and being (or seeming) helpful to peers on the other. Thus, there are important moral restrictions to 'fisherman lies' (Andersen, 1973: p. 159). As Pálsson (1994: p. 914) maintains, 'One way to solve the tricky problem of cooperating under conditions of competition is to participate in an informal club of skippers, a network with relatively stable membership, thereby exchanging information on a regular and reciprocal basis'. In this respect, Barth (1966) refers to 'clusters' or reference groups in which information regarding the proper ways to fish, the location of species and markets, technological and economic innovations is exchanged in the hope that there will be reciprocation in the future. Cross-cutting social ties at community level also mitigate inter-crew rivalry (Byron, 1986).

Nonetheless, fishers thoroughly enjoy the competitive game of being top skippers and top crews. They keep a watchful eye over who lands the most fish. For instance, in Öckerö there is rivalry and competition between crews and a careful keeping of accounts of relative success, prestige and rank. Fishers pay close attention to the latest innovations in boats and gear that will give them an edge over their rivals (Byron, 1993). In view of the competition between boats working out of the same port in Galicia, LiPuma (1992: p. 58) contends that 'the social result is a game of recognition and status whose yardstick is the quantity, quality, and species of fish brought to dock. Fishermen are thus partly defined by this sense of mutual and measurable competition'. In many fishing communities, there are 'leaders' and 'followers'. The former are reputed for their fishing skills and enjoy a high prestige. A skipper's incentive is prestige based on 'conspicuous productivity' (Wadel, 1972; Byron, 1980). With the introduction of quota regimes, part of this penchant for competition has disappeared. As Pálsson (1991) reports with respect to Iceland, there is less secrecy concerning fishing spots and 'quota kings' have replaced 'catch kings'. However, if economic problems are severe, fishermen's solidarity may also be affected. As Vestergaard (1990: p. 29) remarks in regard of fishing in Denmark: 'Knowledge of fishing opportunities becomes so economically valuable that it is tempting to increase secrecy and reserve knowledge for one's own purposes or keep it within narrower circles protected by the increasing use of radio scramblers'.

6. SUCCESS, KNOWLEDGE AND SKILL

Some authors explain differential success by pointing to a combination of engine power, equipment and skipper performance (van der Vlist, 1970), others to skipper-crew interactions (Barth, 1966; Baks and Postel-Coster, 1977), while yet others attribute success to the leadership role and personal skills of skippers (e.g., Byron, 1980; Thorlindsson, 1988). Pálsson and Durrenberger (1990) argue that in Iceland the so-called 'skipper effect' is a folk myth which serves social functions. In their view, success mainly depends on fishing effort, i.e. boat size and number of trips. Though such factors are important so too is the social nature of production. Fishing is very much team work and its outcome depends heavily on the experience, skills, cooperation and coordination of the entire crew. In this respect, one could attribute success partly to a 'crew effect'. As Pálsson (1995) later argued, 'the practical skills of fishing skippers are the result of mutual enskilment, the collective property of a community of practice'. A skipper's prestige is negotiated and maintained at three levels: between skipper and crew; within the local fleet; and among the general public (Pálsson, 1991). In Scotland and Ireland, too, the skipper's performance is the topic of continual public scrutiny (Baks and Postel-Coster, 1977; Peace, 1991). It is not necessarily catching the most fish which earns a skipper his reputation of being a good fisherman. For example, on the Breton island of Houat, being a good fisher is related not only to knowledge ('*avoir le pif*' - 'to have the nose'), but also to 'patience' (the capacity to wait), 'luck' (having good luck, bad luck or no luck) and 'courage' (being 'hardy' or 'rugged'). Strikingly, the women play an important role in deciding who is a good fisher and who is not (Jorion, 1976; 1983). Elsewhere, it may be predominantly retired fishers who keep track of prestige scores. Of course, such evaluations of prestige only have meaning in a local - usually subcultural - context.

Fishers often have a specialised and intimate knowledge of the marine ecosystems they exploit though many are at a disadvantage where it concerns formal education (McGoodwin, 1990; Vestergaard, 1990). With the exploring of new technologies and new markets fishing involves a process of continual learning. While according to some observers and fishers electronic gadgets have led to a degree of deskilling, the traditional method of 'reading' and understanding 'natural signs' continues to be of great importance (Pálsson, 1995; Dufour, 1996). This cultural competence is earned only through experience at sea (Meltzoff and LiPuma, 1986).

But ecological knowledge is no longer sufficient to successfully run a fishing enterprise. In addition, fishers increasingly need to acquire knowledge concerning markets, exchange rates, bureaucracies, political negotiations and so on. Delbos and Prémel (1996: p. 137) argue that 'such skills of anticipation are precisely the modes of operation of the large processing firms and distribution groups which by these means are able to build up their dominant positions and dispossess the producers of their status as social actors'. Small scale, inshore fishers would seem to bear the brunt of this development. They are increasingly marginalised or even ousted from the fishing industry. Capitalisation often results in a shift away from small scale fishermen's control of the means of production. For instance, in the Galician fishing industry, a mode of production based on kinship and community is gradually and partially replaced by capitalist relations of production (LiPuma and Meltzoff, 1989; LiPuma, 1992).

7. THE IDIOM AND IDEOLOGY OF LUCK

Differential success is sometimes explained in terms of such elusive factors like 'luck'. Even when modern technology is used, ideas about luck are often used to explain variations in catches. Clark (1978) maintains that in the English maritime community of Staithes there is a widespread belief in luck, fate, coincidence, and so forth. In Sweden, 'fishing luck' is regarded as an individual resource that can be manipulated by others, for example by 'stealing' luck (Löfgren, 1989). In the Portuguese maritime community of Nazaré 'the importance of knowledge and experience is clearly recognized, but in many ways they fail as explanations. In these cases supernatural explanations are invoked' (Brøgger, 1989). For example, 'bad luck' is sometimes 'explained' by transgression of certain taboos (van Ginkel, 1987). There is indeed a large element of unpredictability and uncertainty in fishing and often bumper catches are explained by reference to luck. But even when production is to some extent controlled, as in shellfish farming, shellfish planters may refer to luck rather than expertise, skill and effort to account for their relative success (van Ginkel, 1994). It is striking that in many places, the concept of luck constitutes the core of the *Weltanschauung* of fishermen: it is 'an important, deeply symbolic, and ubiquitous concept in the consciousness of fishers, used to explain in-group differences in catches' (McGoodwin, 1990: p. 138).

Chance remains an important factor in the spotting and catching of fish. However, some fishers are consistent winners while others are consistently less successful. The notion of 'luck' can function to 'explain' the leadership of successful skippers while playing down differences between crews, avoiding a loss of reputation for less successful skippers and maintaining an egalitarian ideology (Knipe, 1984; Byron, 1986, 1988; Brøgger, 1989; van Ginkel, 1994). By stressing, or even reifying, such elusive factors like luck, potential inter-crew conflicts over catches and income can be avoided and status distinctions mitigated. The ascription of success to chance implies that one does not need to emphasise one's own capabilities if one is successful, and gives those who are unsuccessful an opportunity to attribute this to a lack of fortune rather than personal failure or incompetence. However, reference to 'luck' as an explanation of differential success should not only be understood in functionalist terms, or as key to fishermen's impression management; chance is indeed embedded in the real experiences of fishers and is part and parcel of their day-to-day existence.

8. CONSERVATISM AND ENTREPRENEURIALISM

Given the daily variation in catches and rewards, most fishers have short term time horizons and prefer short term planning (Johnson, 1979). Many have a disinclination for deferred gratification such as savings and investments. It is sometimes reported that fishers are conservative when it comes to values (Cohen, 1987) and innovations (Acheson, 1981). Andersen (1979: p. 21) contends that 'fishermen are ill-positioned to be innovative and progressive, to influence the direction of their fisheries, because they are not in optimal control of their labor, capital, products, and profits'. However, there are marked differences in this respect. On the Dutch island of Texel, fishers from one community were reluctant to adopt new technologies, while fishers from a neighbouring community were among the most innovative in the country. With respect to Shetland fishermen, Byron (1980: p. 231) states: 'Far from being rigid traditionalists, commercial fishermen are realistic businessmen making calculated choices'. If they want to remain in business, presently fishers have to behave as 'modern entrepreneurs'.

Due to the cultural emphasis on independence and individualism, fishers are often suspicious of and resent interference from fisheries policy makers and managers, especially so if measures are believed to be flawed or unjust: 'The motivations of actors in the fisheries and the moral legitimacy of management measures are closely linked with meaning and values' (T. Vestergaard, 1996: p. 87). Underlying fishers' scepticism and distrust *vis-à-vis* the state, civil servants and biologists and the managerial rationalism they embody, is a basic clash of 'scientific' and 'folk' models concerning, among other things, fish stock fluctuations. Fishers see natural processes as dynamic, unpredictable, complicated, disordered, chaotic. In their linear view, scientists (and those who heed their advice) usually depart from the assumption that without human intervention fish stocks are 'ordered, balanced and in dynamic equilibrium'

(Smith 1990: p. 5). In other words, the images of nature held by fishers on the one hand and fisheries scientists, managers and policy makers on the other, and the perceptions and beliefs linked with them, are often at loggerheads. In this respect, too, fishers would seem to be conservative; they often stubbornly resist change in resource management regimes and they have occasionally given vent to their discontent through protests, blockades, non-compliance and confrontations with competitors and external authorities.

From the fishers' point of view, they usually have good reasons to behave like this. New management regimes often impede time-honoured fishing strategies, adaptive performance, flexibility and switching behaviour. For example, individual quota regimes can lead to rigidity in that they create incentives to only seek species one is entitled to catch. To achieve compliance with rules and regulations, these have to be perceived as legitimate and just. What Moustgaard (1996: p. 15) writes about fishers from the Danish island of Bornholm is equally valid in many other cases: 'Like other marginalised groups, the fishermen have gradually developed a set of attitudes constituting the moral basis for breaking the rules, which are perceived as discriminating, unjust, and made up by outsiders'. Devolved management systems, in which fishers are involved in decision making processes concerning the management of marine resources, are an important step in the direction of achieving legitimacy. At the same time, in European fishing communities there is still a strong tendency to accuse 'others' of overfishing and to not accept responsibility for declining fish stocks. Local fishers point an accusing finger in the direction of their compatriots in neighbouring communities; fishers from one nation do likewise with respect to fishers from another nation.

9. COMMUNITY LIFE, IMAGE AND IDENTITY

Fishing permeates non-working life in maritime communities. It receives a good deal of cultural attention and in many instances marks local identity. The communication and reinforcement of norms and values connected with the fishermen's occupation partly occurs in non-working situations. The reputation and ranking of skippers and crews is often evaluated on shore. On the Dutch island of Texel, birthday parties of fishers are often attended by other fisher families and conversations usually focus on fishing. Here, news may be exchanged about the behaviour of other fishers, good fishing locations, new electronic gadgets, and so on. It is therefore important to join social occasions in order to keep up-to-date and gather scanty information. The same goes for attending weddings and funerals, meetings and quayside gatherings. Eating and drinking together ashore is also important to reinforce social relationships, relieve tensions, resolve conflicts and exchange news and information (Peace,

1991; Johnson, 1979). Likewise, participation in rituals such as the Blessing of the Fleet in Roman Catholic countries, and fisheries related prayer and thanksgiving days in Protestant communities in the Netherlands may serve similar purposes. They are further important for maintaining and reinforcing the social fabric of fishing communities.

The distinctive identity related to fishing as an occupation often sets maritime communities apart from other social formations, for example farming communities. In a wider context, the status and prestige ascribed to the occupation of fisherman differs from place to place (Moustgaard, 1984) and from period to period. For example, fishing in eastern Scotland and in Portugal has long been regarded as an undesirable and low prestige occupation (Nadel-Klein, 1988; Cole, 1991), whereas in the Netherlands, fishers used to be considered hardworking, deeply religious 'noble commoners' (an image which has faded rapidly over the past few decades). More generally, however, fishers tend to be 'a denigrated if not despised segment' of the societies in which they live (Smith, 1977: p. 8), and in Europe, they have long been viewed as belonging to the lowest social classes and stigmatised accordingly (Coull, 1972). Intermarriage between landowners and fishers was usually rare, and many fishing villages were characterised by endogamous marriage patterns (Cole, 1991; LiPuma, 1992; van Ginkel, 1993).

But fishers may use social stigma to express, articulate and cherish their identity as a fishing community. For instance, Cole (1991: p. 46) reports that the *pescadores* of Vila Chã, a Portuguese fishing village 'have chosen to value as the foundation of a cherished group identity the very areas upon which their social stigmatization is based - the organization of maritime production, the gendered division of labor within the household, and the construction of social images of maritime women, for example'. The villagers express their pride in being (descendents of) *pescadores*, even those who are currently no longer engaged in fishing. Likewise, Nadel (1984: p. 113) states that in Ferryden, a Scottish 'maritime' community few of whose inhabitants have spent any time fishing, 'fisherfolk identity has become a cherished posession, highly romanticized and mythologized'. The Ferrydeners cling to their image of themselves as 'fishers'. In both Ferryden and Vila Chã, there is symbolic boundary marking and maintenance by means of which fishers or descendents of fishers distinguish themselves from the social world of 'landlubbers' around them and to award themselves special status. In Vila Chã, such boundary maintaining mechanisms 'include the operation of an informal network for the distribution of fresh fish; a profession of knowledge and interest in things maritime, like the weather, the boats, and tales of former fishing days that express their attachment to the sea; the attribution of special virtues, such as hospitality, to the pescadores ...; and the use of personal nicknames that have meaning only for insiders, only for members of the group' (Cole, 1991: p. 46).

10. CONCLUSION

This brief description of cultural aspects of fishing and fishing communities cannot do justice to the many intricacies of 'real life' in Europe's economically, socially and culturally widely diverse fishing settings. Nonetheless, it covers some themes that seem to be common in many such social formations. In addition, fishing communities throughout Europe have faced similar economic and socio-cultural consequences of modernisation and national and supranational interventions in fisheries management. These include: increasing antagonism and conflicts among fishers and between fishers and state representatives; growing inequality and social divisions; a consolidation of vested interests; the blocking of social mobility; division and opposition between boat owners and crewmen; heavy indebtedness and bankruptcies; a decline in the status of fishers; a loss of certain fishing skills; destruction of communal identity as a fishing culture; changes in community values and association; and a loss of autonomy. In many European coastal communities, fishing no longer dominates 'local culture' but has turned into an occupational subculture (Löfgren, 1984).

Inshore fishers seem to be pessimistic about the future, and for good reason; the number of inshore fishers has been on the decline continually for several decades in most of Europe's fishing communities. In many places, young people are increasingly reluctant to follow in their father's footsteps. In view of the 'breakdown' of Danish fishermen's social identity, Vestergaard (1990: p. 29) even contends that they 'have come to feel like a dying species'. What is particularly worrying is that, given the fact that many communities depend on fishing and that entire local families are involved in the fishing industry in one way or another, a crisis in the fisheries potentially undermines the entire economic basis and social fabric of such social formations.

But given the feelings of pride and satisfaction they derive from their occupation, many fishers tenaciously adhere to it even when facing declining stocks and catches, and substandard incomes. Their relation to fishing is expressive and existential: to them, fishing is a lifestyle and they greatly value the idea of 'autonomy' and the corresponding identity as 'independent individualists'. Therefore, losing their jobs as fishers would mean much more than losing their livelihood. Fishing permeates their entire personas and their image of themselves. For many, giving up as a fisher would mean giving up their dignity and pride. Therefore, fishers often persist in working in a failed fishery. Moreover, they are ill prepared for land based jobs. Their formal education is often poor and the knowledge and skills learned in fishing are of little or no use on shore. Nonetheless, communities dependent on inshore fishing and consisting of simple commodity producers attempt to 'muddle through' in adverse times. They have proven to be extremely resilient and adaptive in the past, but today they must cope with forces that are beyond their understanding and control.

11. REFERENCES

Acheson, J.M. (1981) Anthropology of fishing, *Annual Review of Anthropology*, **10**, 275-316.

Alegret, J.-L. (1996) Ancient institutions confronting change: the Catalan fishermen's cofradías, in K. Crean and D. Symes (eds.), *Fisheries Management in Crisis*, Blackwell Science, Oxford, pp. 92-98.

Alegret, J.-L. (1998) Property rights, regulating measures and strategic responses among the fishermen of Catalonia, in D. Symes (ed.), *Property Rights and Regulatory Systems in Fisheries*, Blackwell Science, Oxford, pp. 175-187.

Andersen, R. (1973) Those fisherman lies: custom and competition in North Atlantic fisherman communication, *Ethnos*, **38** (1-4), 153-164.

Andersen, R. (1979) Introduction: North Atlantic maritime peoples in transition, in R. Andersen (ed.), *North Atlantic Maritime Cultures. Anthropological Essays on Changing Adaptations*, Mouton, The Hague, pp. 1-28.

Baks, C. and Postel-Coster, E. (1977) Fishing communities on the Scottish East Coast: traditions in a modern setting, in M.E. Smith (ed.), *Those Who Live from the Sea. A Study in Maritime Anthropology*, West, St. Paul, pp. 23-40.

Barth, F. (1966) Models of social organization. *Royal Anthropological Institute Occasional Paper*, **23**, London.

Brøgger, J. (1989) *Pre-bureaucratic Europeans. A Study of a Portuguese Fishing Community*, Norwegian University Press, Oslo.

Byron, R.F. (1975) Economic functions of kinship values in family businesses: fishing crews in North Atlantic communities, *Sociology and Social Research*, **60** (2), 147-160.

Byron, R.F. (1978) Unequal equals: the occupational culture of Shetland fishermen. Paper presented at the *SSRC Seminar on Anthropological Research in Western Europe*, University of Sussex, 16-17 December 1978.

Byron, R.F. (1980) Skippers and strategies. Leadership and innovation in Shetland fishing crews, *Human Organization*, **39** (3), 227-232.

Byron, R.F. (1986) *Sea Change. A Shetland Society, 1970-79*, ISER, Memorial University of Newfoundland, St. John's.

Byron, R.F. (1988) Luck and leadership: the management of decisions in Shetland fishing crews, *Maritime Anthropological Studies*, **1** (1), 3-14.

Byron, R.F. (1993) Öckerö: domestic organisation and social change in a Swedish maritime community, *Ethnologia Scandinavica*, **23**, 58-72.

Byron, R.F. (1994) The maritime household in Northern Europe, *Comparative Studies in Society and History*, **36** (2), 271-292.

Cohen, A.P. (1987) *Whalsay. Symbol, Segment and Boundary in a Shetland Island Community*, Manchester University Press, Manchester.

Clark, D. (1978) Religion and superstition in a fishing community, *Research Bulletin*, Institute for the Study of Worship and Religious Architecture, University of Birmingham, 59-69.

Cole, S. (1991) *Women of the Praia. Work and Lives in a Portuguese Coastal Community*, Princeton University Press, Princeton.

Coull, J.R. (1972) *The Fisheries of Europe. An Economic Geography*, G. Bell and Sons, London.

Delbos, G. and Prémel, G. (1996) The Breton fishing crisis of the 1990s: local society in the throes of enforced change, in K. Crean and D. Symes (eds.), *Fisheries Management in Crisis*, Blackwell Science, Oxford, pp. 129-140.

Dufour, A-H. (1996) Management and practice in the small-scale inshore fisheries of the French Mediterranean, in K. Crean and D. Symes (eds.), *Fisheries Management in Crisis*, Blackwell Science, Oxford, pp. 110-115.

Frangoudes, K. (1996) Fishermen households and fishing communities in Greece: A case study of Nea Michaniona, in K. Crean and D. Symes (eds.), *Fisheries Management in Crisis*, Blackwell Science, Oxford, pp. 119-128.

Hanna, S. and Jentoft, S. (1996) Human use of the natural environment, in S.S. Hanna, C. Folke and K.-G. Maler (eds.), *Rights to Nature: Ecological, Economic, Cultural and Political Principles of Institutions for the Environment*, Island Press, Washington, DC, pp. 35-55.

Jentoft, S. and Davis, A. (1993) Self and sacrifice. An investigation of small boat fisher individualism and its implication for producer cooperatives, *Human Organization*, **52** (4), 356-367.

Jentoft, S. and Kristoffersen, T. (1989) Fishermen's co-management: the case of the Lofoten fishery, *Human Organization*, **48** (4), 355-365.

Johnson, T. (1979) Work together, eat together: conflict and conflict management in a Portuguese fishing village, in R. Andersen (ed.), *North Atlantic Maritime Cultures. Anthropological Essays on Changing Adaptations*, Mouton, The Hague, pp. 241-252.

Jorion, P. (1976) To be a good fisherman you do not need any fish, *Cambridge Anthropology*, **3**, 1-12.

Jorion, P. (1978) Marks and rabbit furs: location and sharing of grounds in coastal fishing, *Peasant Studies*, **7** (2), 86-100.

Jorion, P. (1983) *Les Pêcheurs d'Houat. Anthropologie Économique*, Hermann, Paris.

Knipe, E. (1984) *Gamrie: An Exploration in Cultural Ecology. A Study of Maritime Adaptations in a Scottish Fishing Village*, University Press of America, Lanham.

LiPuma, E. (1992) Social identity and the European Community. An Iberian example, *Maritime Anthropological Studies*, **5** (2), 46-73.

LiPuma, E. and Meltzoff, S.K. (1989) Toward a theory of culture and class: an Iberian example, *American Ethnologist*, **16** (2), 313-334.

Löfgren, O. (1972) Resource management and family firms: Swedish west coast fishermen, in R. Andersen and C. Wadel (eds.), *North Atlantic Fishermen: Anthropological Essays on Modern Fishing*, ISER, Memorial University of Newfoundland, St. John's, pp. 82-103.

Löfgren, O. (1977) Maritime hunters in industrial society. The transformation of a Swedish fishing community 1800-1970, in O. Löfgren (ed.), *Fångstmän i Industri Samhället*, LiberLäromedel, Lund, pp. 226-235.

Löfgren, O. (1984) The making of a fisherman. The social context of socialization in a Swedish fishing community, in B. Gunda (ed.), *The Fishing Culture of the World*, Akadémiai Kiadó, Budapest, pp. 281-296.

Löfgren, O. (1989) The reluctant competitors: fisherman's luck in two Swedish maritime settings, *Maritime Anthropological Studies*, **2** (1), 34-58.

McGoodwin, J.R. (1990) *Crisis in the World's Fisheries. People, Problems, and Policies*, Stanford University Press, Stanford.

Meltzoff, S.K. and LiPuma, E. (1986) The troubled seas of Spanish fishermen: marine policy and the economy of change, *American Ethnologist*, **13** (4), 681-699.

Mondardini Morelli, G. (1998) Traditional use rights in the small-scale fisheries and marine protected areas between Sardinia and Corsica, in D. Symes (ed.), *Property Rights and Regulatory Systems in Fisheries*, Blackwell Science, Oxford, pp. 188-200.

Moustgaard, P. (1984) The fishing community, the gear and the environment, in B. Gunda (ed.), *The Fishing Culture of the World*, Akadémiai Kiadó, Budapest, pp. 337-357.

Moustgaard, P. (1996) In the place of God: fishermen, regulations, and the Baltic cod. A case study from the Danish island of Bornholm, *MARE Working Paper* No. 5, Aarhus University, Aarhus.

Nadel, J. (1984) Stigma and separation: pariah status and community persistence in a Scottish fishing village, *Ethnology*, **23** (2), 101-115.

Nadel-Klein, J. (1988) A fisher laddie needs a fisher lassie: endogamy and work in a Scottish fishing village, in J. Nadel-Klein and D.L. Davis (eds.), *To Work and to Weep. Women in Fishing Economies*, ISER, Memorial University of Newfoundland, St. John's, pp. 190-214.

Nadel-Klein, J. and Davis, D.L. (eds.) (1988) *To Work and to Weep. Women in Fishing Economies*, ISER, Memorial University of Newfoundland, St. John's.

Pálsson, G. (1991) *Coastal Economies, Cultural Accounts. Human Ecology and Icelandic Discourse*, Manchester University Press, Manchester.

Pálsson, G. (1994) Enskilment at sea, *Man* N.S., **29** (4), 901-927.

Pálsson, G. (1995) Fisheries, in M. McDonald (ed.), *Towards an Anthropology of European Union*, Report to the European Commission, s.p. European Commission (S.G.), Brussels.

Pálsson, G. and Durrenberger, E.P. (1990) Systems of production and social discourse: the skipper effect revisited, *American Anthropologist*, **92** (1), 130-141.

Peace, A. (1991) Fishing, drinking and the construction of identity in rural Ireland, *Maritime Anthropological Studies*, **4** (2), 3-16.

Peace, A. (1996) When the salmon comes: the politics of summer fishing in an Irish community, *Journal of Anthropological Research*, **52** (1), 85-106.

Prattis, J.I. (1973) A model of shipboard interaction on a Hebridean fishing vessel, *Journal of Anthropological Research*, **29** (3), 210-219.

Rasmussen, A.H. (1974) The development and decline of the coastal fishery on the Danish West Coast, *Ethnologia Scandinavica*, **4**, 117-133.

Smith, M.E. (1977) Introduction, in M.E. Smith (ed.), *Those Who Live From the Sea. A Study in Maritime Anthropology*, West, St. Paul, pp. 1-22.

Smith, M.E. (1990) Chaos in fisheries management, *Maritime Anthropological Studies*, **3** (2), 1-13.

Taylor, L.J. (1977) *'Teilion an Éisc' 'Teelin of the Fish': Fishing and Community in Southwest Donegal*, Ph.D dissertation, State University of New York, Stony Brook, N.Y.

Taylor, L.J. (1981) 'Man the fisher': salmon farming and the expression of community in an Irish settlement, *American Ethnologist*, **8** (4), 774-788.

Thorlindsson, T. (1988) The skipper effect in the Icelandic herring fishery, *Human Organization*, **47** (3), 199-212.

Trice, H.M. (1993) *Occupational Subcultures in the Workplace*, ILR Press, Ithaca.

van der Vlist, R. (1970) *Verschillen in Groepsprestaties in de Nederlandse Zeevisserij*, Wolters-Noordhoff, Groningen.

van Ginkel, R. (1987) Pigs, priests and other puzzles: fishermen's taboos in anthropological perspective, *Ethnologia Europaea*, **17** (1), 57-68.

van Ginkel, R. (1993) *Tussen Scylla en Charybdis. Een Etnohistorie van Texels Vissersvolk (1813-1932)*, Het Spinhuis, Amsterdam.

van Ginkel, R. (1994) 'One drop of luck weighs more than a bucketful of wisdom'. Success and the idiom and ideology of Dutch shellfish planters, *Ethnologia Europaea*, **24** (2), 155-166.

van Ginkel, R. (1996) Cooperating competitors. Texel fishermen and their organizations, *Anthropological Quarterly*, **69** (2), 51-65.

van Maanen, J. and Barley, S.R. (1984) Occupational communities, *Research in Organizational Behavior*, **6**, 287-365.

Vestergaard, E. (1996) Voices from the fishermen. Views on regulations, biologists and the state, *MARE Working Paper* No. 6, Aarhus University, Aarhus.

Vestergaard, T. (1990) The fishermen and the nation: the identity of a Danish occupational group, *Maritime Anthropological Studies*, **3** (2), 14-34.

Vestergaard, T. (1996) Social adaptations to a fluctuating resource, in K. Crean and D. Symes (eds.), *Fisheries Management in Crisis*, Blackwell Science, Oxford, pp. 87-91.

Wadel, C. (1972) Capitalization and ownership: the persistence of fishermen-ownership in the Norwegian herring fishery, in R. Andersen and C. Wadel (eds.), *North Atlantic Fishermen. Anthropological Essays on Modern Fishing*, ISER, Memorial University of Newfoundland, St. John's, pp. 104-119.

Chapter 11

AN ECONOMIC RATIONALE FOR INSHORE FISHING: SIMPLE COMMODITY PRODUCTION AND THE LIFE MODE APPROACH

Kirsten Monrad Hansen and Thomas Højrup

1. INTRODUCTION

Earlier chapters have uncovered the great diversity of activities that are subsumed under the term inshore fisheries and so exposed the basic dilemmas surrounding any attempt to generate a simple, comprehensive definition. Inshore fisheries express themselves in many different forms: they include conventional capture fisheries, which may be undertaken commercially on a full time, part time or seasonal basis or for purposes of recreation, and culture fisheries almost invariably conducted on a commercial basis. Moreover, inshore fisheries are not dominated by the conventional capitalist division between investors, companies and employees. European rural society reveals both a simple commodity mode of production and a capitalist mode of production often existing side by side (Højrup, 1983). Each contains a number of logical variants which co-exist in differing degrees of interdependence and opposition. Under various forms of finance capital or state planning, private, competition oriented, entrepreneurial capitalism coexists with monopoly capitalism, state capitalism and simple commodity production, with the latter organised either as family based enterprises, as for example the family farm, partnership companies, most commonly found in professional, service and manufacturing activities, or in joint or share ownership as in fishing. In fact, practically all forms of organisation associated with capitalist and simple commodity production can be found within the inshore fisheries sector in Europe.

D. Symes and J. Phillipson (eds.), Inshore Fisheries Management, 195–205.

It would be naive, therefore, to pretend that a single economic rationale would suffice for the inshore sector or to imagine that inshore fisheries are in some way exempt from the economics of the firm, the rigours of the market place and the tyranny of profit and loss accounts. What may well be true, however, is that some organisational forms will respond in very different ways to conventional economic signals and adopt markedly different forms of survival strategies.

Several macro-economic trends have influenced the profitability of inshore fisheries over recent years. One of the most significant has been the effects of the globalisation of the food industry and the downward pressure on quayside prices for fish caused by the ever widening sources of supply of both fish and competing foods for the urban markets. In general the industry has been unable to offset these developments by increasing the volume of production; fish stocks are depleted and, in any case, increasing supplies to the market would simply depress prices further. The exception to this rule has occurred in the shellfish sector which has responded positively to the growth in demand for its high priced, luxury products.

At the same time, there has been a significant shift in the relative costs of capital and labour affecting economic production in general, the fishing industry in particular and the inshore sector only to a slightly less marked extent. Due to the development and near universal diffusion of new technologies, coupled with the availability of grant aid and low interest loan schemes for modernisation projects, the relative price of capital has been falling. Meanwhile, as a consequence of widespread wage inflation and rising social insurance charges, the price of labour has been rising. As a result - and as noted elsewhere in this volume, but especially in Chapter 9 - the cost efficiencies of substituting capital for labour have impacted quite sharply on the inshore sector. New technologies have also encouraged changes in the levels of cost efficiency between different fishing methods deployed in the inshore sector. Very few technological innovations have failed to percolate through to the small scale, inshore fisheries in some form or other.

Finally, a major concern for the traditional inshore fisheries in many parts of Europe is the increasing external costs imposed by the expansion of recreational fishing and the explosive development of marine aquaculture (see Chapter 13). Both have tended to challenge the commercial capture fisheries in the market place. In the case of recreational fisheries, the challenge comes from the growth of a 'grey market' for the informal sales of fish to local trade outlets (shops and restaurants) as well as to friends and neighbours. There is also a problem with the burgeoning activity of ferrying recreational fishermen out to the fishing grounds to jig for cod, which disturbs the commercial gillnet fishery partly on account of the significant loss of lures but also because of the considerable catches of cod made in some areas. A further source of external costs which could impact quite heavily in the future is the burden of restrictions on fishing activity especially in inshore waters in pursuit of more environmentally friendly fishing practices (see Chapter 14) - these restrictions could involve the further designation of marine protected areas and the banning of technically efficient but 'inappropriate' fishing gears.

Against a background of generally rising costs and stable, falling or at best uncertain revenues, the decision to enter the fishing industry is an increasingly difficult one, especially in cases where entry implies high initial costs of acquiring both boat and licence or where remaining in the industry involves the renewal of capital through the modernisation of the boat and the replacement of fishing gear. The decision is made all the more difficult by the likelihood of privatisation of fishing rights in the form of individual transferable quotas (ITQs) for certain species and/or the transfer of part of the management costs to the individual fishermen in the form of licence fees. Access to the fisheries these days is neither open nor free and the days when the inshore fisheries in particular afforded some kind of buffer against economic recession and unemployment in Europe's coastal regions are long gone. Instead, inshore fishermen are themselves required to adjust their economic behaviour to take account of rising costs, increasing regulation and low incomes.

2. SIMPLE COMMODITY PRODUCTION

In comparison with capitalist firms, simple commodity production, where small scale self-employed producers own their own means of production, is frequently characterised as being especially resistant to market fluctuations. It is capable of functioning for quite long periods without earning revenues commensurate with the value of the plant and equipment involved in production - commensurate in the sense of matching the levels required to sustain a capitalist business in production. The owner operator's income can often be supplemented by earnings from ancillary activities which help to keep the enterprise afloat or maintain plant and equipment in working order during periods of poor landings or weak prices. Moreover, a persistent feature of simple commodity production is the undertaking of several functions, normally discharged to third parties, within the production unit itself. The skipper-owner, for example, may be engaged in the direct selling of the catch, together with the repair and maintenance of the boat and fishing gear - activities which both extend his competencies and also internalise some of the current costs of production. Such enterprises can therefore survive under conditions where business capital might be withdrawn from production and invested in other sectors. The resilience of simple commodity production means that unless especially favoured by economies of scale, capitalist businesses are usually unable to oust small scale producers. Their best hope is for the indirect economic subordination of simple commodity producers through the need to obtain loans to tide them over periods of economic difficulty.

It is commonly assumed that technological innovations tend to benefit the larger, capital intensive enterprises and reinforce the economies of scale. This may be true initially, but the adaptation of the innovation and its absorption by the artisanal sector can eventually dissipate some of the advantages held by large scale operators. The diesel engine, electronic navigation and fish finding equipment are all examples of major technological developments that have ultimately narrowed, rather than widened, the advantages of large scale production.

In agriculture, segments of the sector which benefit most from economies of scale have generally been transferred to specialised secondary industries appropriated by the farmers themselves, while farm production has remained subject to continual improvements in productivity, matching similar developments in much larger economic units. The same is broadly true of the fishing industry. Fishing in coastal waters throughout northern and western Europe is based upon resources which are quite widely dispersed. Hence the advantages inherent in large scale operations such as segmentation and specialisation of the fleet into large trawlers with attendant carrier vessels - as associated with offshore pelagic fisheries - can offer little by way of increases in productivity. There is, in fact, no advantage in large scale specialisation, since the real efficiency in production in inshore fishing and the satisfaction of market demand for good quality, fresh fish calls for fishing units to be fast, flexible, suited to harvesting smaller concentrations of fish and able to switch between different species and different gears. In these circumstances, fishing undertaken with small, skipper owned boats and equipped with state of the art technology is the most competitive form of fishing for the fresh fish market.

3. THE LIFE MODE OF THE SELF-EMPLOYED

Whereas the wage system may be said to provide an organisational structure for a capitalist enterprise, the core of simple commodity production is not organised around economic relations but rather on the basis of social relations in the form of family ties, cooperative relations between partners and other ideologically based associations which bind producers together into cohesive production units. Joint ownership and share systems in both fishing and small scale farming serve the same functions. As a result the concepts of 'partners' and 'family' assume a radically different significance than among wage earners for whom family signifies a framework for socially reproductive activities rather than for economic relations.

In socially constructed partnerships or family enterprises, the prime concern may simply be to maintain production. If consolidation of the enterprise appears feasible, then large investments may well be made, diversification pursued or ancillary activities undertaken in order to supplement income. But one can never be sure whether these enterprises are producing to full capacity for two reasons: first, because it is not absolutely necessary to maximise revenues in line with interest payments on the capital value of the plant and equipment beyond that needed for replacement and to keep abreast of developments in productivity. For this reason, conventional economic analysis has difficulty in identifying appropriate measures of economic performance in the small boat sector (Boncoeur *et al.*, 2000). Secondly, it is difficult to estimate the level of latent productive capacity which remains hidden in the particular social network which defines the enterprise. The goal of the partnership or family enterprise may ultimately be to remain self-employed. It is, therefore, a means that serves its own end.

Because of this goal, simple commodity production is fundamentally different from capitalist business. Even though there is nothing to prevent a simple commodity producer earning more than is necessary to keep the wheels of production turning, there is no mechanism to ensure that the enterprise will produce at an optimal level, seen from a strictly financial point of view. For this reason, the 'law of value' does not directly apply. A fall in price below the cost of producing the catch is not automatically countered by a decline in production brought about through the withdrawal of some of the producers. Confronted by falling prices, simple commodity producers will instead attempt to increase their output, often in fisheries by diversifying into other locally available species, and if need be supplement their revenues and subsidise the fishing enterprise through undertaking additional activities or even wage work. In the familiar work combination of farming and fishing, which exploited the same or at least overlapping social networks, it was difficult if not impossible to separate the economics of fishing from the economic performance of the farm. Cross-subsidisation of these joint activities was an endemic feature of the traditional pluriactive economy.

The notion of work therefore assumes a very different cultural content than it does in a wage earning economy. A life mode based on simple commodity production makes no distinction between 'work time' and 'free time', which are essential antitheses in a capitalist work environment. Free time has no real meaning: you are never free from work because you are never put to work. Instead, you put yourself to work; you involve yourself in work, because this involvement is the prerequisite for and, indeed, the essence of being self-employed. You remain responsible for the operation and the success of the enterprise and you are liable for its failings. Although inevitably subject to some degree of social obligation in relation to other partners in the network, this attitude to work is the precondition for being able to perceive all the partners as free and independent: the partners work together out of a common sense of purpose and inner drive. Significantly, as is made clear in Chapters 9 and 10, small boat crews tend not to recognise any differences in status on board the boat. Work and the rewards for work are shared equally - a very different situation to that encountered on the much larger, company owned and industrially structured offshore trawlers.

4. THE LIFE MODES OF EMPLOYEES

Unlike the simple commodity producer, wage earning workers in the capitalist system of production possess neither the means of production nor the qualifications needed to set in motion and control the entire production process. Unskilled workers are incorporated into a long, complex and often tedious production process, wherein their individual tasks are commonly confined to a single phase of the production sequence: the 'skills' are often internalised within the machinery itself. Alternatively, qualifications may be demanded of the employee where the task is to manage more complex machines; the modern skilled worker becomes involved with a sophisticated production process and may be responsible in part for the quality of the finished product.

For the employee work serves the function of providing an income which makes it possible to live a meaningful life beyond work. Work is merely a means of earning free time and is carried out not for its own sake but with the sole aim of achieving a wage. But wages are not simply a product of the work done - they must be negotiated, usually on a collective rather than individual basis and through unions which in the past have sought to monopolise the supply of labour within a particular sector of the production process. The resulting wage levels are, in fact, the product of a complex set of factors many of which are external to the actual work performed. The exchange value of labour power is thus culturally, rather than economically, determined. By contrast, in simple commodity production there is no such commodity as labour: labour, or more properly work, is embedded in the skipper-owner or share system as life mode. In the case of the wage earner one does not work for the sake of the business and the enterprise's competitiveness is of little direct concern, except perhaps where profit sharing schemes are in place - as is commonly the case in the industrially structured offshore trawling sector.

On the other hand, in a career oriented life mode one strives to make demands on oneself by fulfilling the expectations of leadership, by improving qualifications, by involvement in additional tasks and by surpassing colleagues. Here the concept of work stands in sharp contrast to the ideas of labour or worker solidarity inherent in the unskilled or semi-skilled wage earner's life mode. The career professional does not attain greater freedom by becoming self-employed or by earning a higher wage but through advancement and a change in status within the firm. The concepts of freedom and involvement are wholly related to the work situation and to a career perspective. For the professional the division between work and free time may become just as illusory as in simple commodity production: he may choose to use the free time to develop personal and professional skills and cultivate relationships as an aid to further career advancement. But unlike the simple commodity producer only very rarely will the career professional actually own the means of production. Nor will he recognise any compelling social obligations which tie him to a particular local social network.

5. SIMPLE COMMODITY PRODUCTION AND THE PRINCIPLE OF OPEN ACCESS

In western Europe the capitalist and simple commodity modes of production have coexisted for hundreds of years in the fishing industry. The capitalist mode dominated the high seas fisheries from the fifteenth century onwards. But in the last century small scale fishing units extended their field of operation at the expense of the old company controlled vessels, thus transforming areas like the North Sea from a zone of middle-water fisheries to one suitable for exploitation by a modernised small boat fleet.

Similarities in the differentiation of fishermen in most European countries are partly related to the different life modes outlined above and also to the ways in which resource management systems impact on fishing behaviour. The principle of open access, which addresses the crucial question of access to fishing rights, offers one means of explaining the underlying differentiation and the basic similarities found across Europe. Three levels of specification of the principle can be identified. In the first place, the state may decide to prevent fisheries becoming subject to the full privatisation of fishing rights, thus nominally allowing all fishermen free access to the resource. Secondly, to prevent conflicts arising in the harvesting of the same stock by boats using different gears, common rules are typically set down to regulate the way in which fishing units relate to each other while on the fishing grounds, including rules for separating different gears in time and/or space which prevent the overlap of incompatible fishing gears. The rules governing the seasonal Lofoten cod fisheries are an outstanding example of this approach. Finally, when fishermen using different harvesting methods clash over how the ecosystem should be exploited - accusing each other of promoting the overfishing of the stocks - the state is required to arbitrate through the introduction of technical conservation measures and/or quota restrictions (Hansen, 1997). In all three instances the state's actions are intended to be neutral, though there is a general belief that most state-led management regimes - and especially those based on quota regulation - will tend to favour the interests of capital over those of the individual small scale enterprise.

Both modes of production can remain viable under these conditions. No doubt the owners of capital would prefer the privatisation and trading of fishing rights because it makes possible the deployment of capital to purchase fishing rights and so eventually monopolise the object of labour. The power of the market is used to force simple commodity producers out of business. But where the regulatory system seeks to protect access rights, then both modes of production can coexist and, in the fullness of time, the self-employed skipper owner may prove the more durable.

The simple commodity mode of production is developed on the basis of a commodity market alone. Costs of production are divided into two elements: basic costs or overheads, which correspond to the operating equipment (boat and fishing gears) which determines the unit's production capability, and 'unit costs' (fuel, ice, bait and labour) which vary according to the duration of the trip. Costs are clearly not a linear function of the quantity of fish landed - and the relationship between catch quantity and costs will vary. To a degree, the relationship is defined by the nature of the equipment (i.e. the overhead costs) which can be manipulated to increase production, improve quality or reduce unit costs. The aim of the operation is to ensure that the earnings of the enterprise are at least equal to the expenses incurred during the production cycle.

This balance can be achieved in different ways and finding the appropriate means of balancing the accounts is the central concern of the simple commodity producer. There are in effect four components which can be manipulated to create a favourable balance: catch volume, quality of the product which helps to decide the first hand sale price, overheads and unit costs. The key variable for the simple commodity producer has tended in the past to be catch volume. If price is beyond his control, the simplest answer is to increase the volume of the catch; the main concern is to ensure that the catch volume remains above the level needed to cover the costs. With this simple prerequisite in mind, there are four rather different sets of reasoning open to the simple commodity producer.

(i) *Stay at sea just as long as it takes.* The first and most universal reasoning is to spend as much time at sea as is needed to harvest sufficient fish to cover the costs of production; when that level of harvest has been reached fishing can be suspended. In Thorupstrand in the north west of the Danish Jutland peninsula, there are 19 fishing boats varying in size and fishing methods, whose actions seem to confirm this kind of reasoning. There are many days with good fishing weather when the boats remain in harbour. Why? One of the skippers is also president of the local fishermen's association; one has a string of horses; a third has a neat, well cared for garden; and several are Jehovah's Witnesses. In the summer of 1997 one vessel was hauled up on the beach and almost buried in the sand for a month and a half of good fishing - the skipper was on holiday with his family. Only when fish prices fall will it be necessary to spend more time at sea.

(ii) *When prices fall reinvest in the boat and its equipment* with the aim of increasing catch volume or reducing unit costs. One can invest in order to save on fuel, ice or a crew member. With a new engine one can either extend the area of fishing or reach the favoured fishing ground more quickly. With automatic technology for baiting and/or setting and hauling the lines, one can cut out a crew member and so save on unit costs. And when it is not possible to catch more fish because of quota restrictions, it may still be possible to improve methods of handling the fish on board so as to deliver a higher quality product to the quayside and command a higher price. This form of reasoning coincidentally draws attention to the dilemma between increasing the volume of production and improving the quality of the product.

(iii) *When prices are expected to remain low, reassess the investment priorities*. Investment in equipment must not be allowed to generate expenses that exceed the unit's ability to cover overheads and unit costs. It may be necessary to abandon plans to build a new vessel and instead make do with renovating the existing boat and reducing overheads. Fishermen in the northern part of Bohuslan on Sweden's west coast in the late 1960s and early 1970s continued to fish with old but constantly modernised boats at very low overall cost. In the south, the fishermen had invested in large new trawlers. Each took shares in the others' boats or acted as guarantors. But shortly afterwards the target North Sea herring fishery collapsed and it was no longer possible to catch enough fish to cover their basic costs. The new vessels were sold on often at prices well below the original cost of construction and several coastal communities endured a period of considerable hardship. In the north, however, the more cautious approach guaranteed the survival of most enterprises.

(iv) *Develop a more flexible structure to the enterprise* in which the vessel, its equipment and manpower are jointly capable of generating a level of income which, in almost any foreseen circumstances, will be sufficient to cover basic overhead costs. The modern emphasis in fleet management focuses on the ability of self-employed and share fishermen to mould a flexible and dynamic organisation capable of reallocating the basic components (boats, gears and crews) into new patterns as the circumstances dictate. It is this plasticity which explains why some share based enterprises involve only temporary permutations of the basic components - and it is this that serves to distinguish share fishing from the more stable but less flexible production units such as are found in family farming and which tend to retain their individual identities over several generations. The very flexibility of the simple commodity producer is often sufficient reason to guarantee his survival.

6. CONCLUSIONS

The underlying principle for the survival of the capitalist mode of production is that capital is mobile. It can be withdrawn from enterprises exhibiting low profits and reinvested in other businesses which are more profitable. In general capital tends to follow those branches of the fishing industry where economies of scale are likely to guarantee profits and which demand high inputs of capital to assure the viability. On the other hand, simple commodity production, which characterises much - but by no means all - of the inshore fishing sector in Europe, generates its own advantages through the principle of flexible reorganisation outlined in the fourth scenario of the previous section. All the essential components of the system are 'mobile' and can be combined in a way which gives the skipper owned fleet greater adaptability and enables it to outmanoeuvre the much less flexible company owned, large boat fleets in areas like the North Sea.

Systems of property rights and management regimes are of considerable importance in creating the conditions which make it possible for certain life modes to survive, while others are forced to retreat from the inshore fisheries. The survival of viable inshore fishing communities, moulded by a life mode of self-employment rather than wage labour, is closely related to the choice of management regime. Privatisation of fishing rights in pursuit of conventional economic notions of efficiency, involving the allocation of individual transferable quotas and the creation of a free market in quotas, could alter the current situation quite markedly and undermine some of the advantages currently enjoyed by the skipper owned small boat fleet. If one wishes to underpin the economic and social welfare of many of Europe's fishing dependent regions, and at the same time maintain a productive inshore fishing industry able to deliver high quality food products, then it seems sensible to select a management regime which best suits the simple commodity mode of production.

7. REFERENCES

Boncoeur, J., Coglan, L., Le Gallic, B. and Pascoe, S. (2000) On the (ir)relevance of rates of return measures of economic performance to small boats, *Fisheries Research*, **49** (2), 105-115.

Hansen, K.M. (1997) *Fishing and the Environment - Conflict or Harmony?* The Nordic Fishermen's Environmental Secretariat, Nordic Council of Ministers, Esbjerg.

Højrup, T. (1983) The concept of life-mode. A form-specifying mode of analysis applied to contemporary western Europe, *Ethnologia Scandinavica*, 1983, 15-50.

Chapter 12

INSTITUTIONAL ORGANISATION AND REGULATORY SYSTEMS: LOCALITY VERSUS CENTRE DRIVEN APPROACHES TO INSHORE FISHERIES MANAGEMENT

Jeremy Phillipson and Mireille Thom

1. REGULATORY REGIMES

The regulatory regime governing inshore fisheries embraces aspects of institutional organisation - comprising organisational forms, institutional linkages and decision making procedures - together with the measures and approaches utilised for the implementation, monitoring and enforcement of management policy. These components represent some of the key determining factors in the success or failure of fisheries management.

This chapter reflects on the regulatory systems applied to, and the main organisational actors and structures responsible for, inshore fisheries management, and more particularly the choice between central and local approaches to regulation. It borrows heavily from material presented in the earlier national chapters. It is therefore informed by existing experiences of management in north west Europe, but does not attempt to provide an all-embracing or in-depth study of the approaches and issues nor of the milieu of organisations which have an interest or role in inshore fisheries and the coastal zone. Its focus is upon commercial sea fisheries as opposed to recreational fishing and aquaculture.

Deconstructing and analysing the regulatory regime which forms the basis of inshore management is a demanding task. The regime involves structures and provisions which are spread across the full spectrum of actors and levels of organisation within a state and is not necessarily confined to the locality or to locally based approaches. Regulatory regimes and legal

D. Symes and J. Phillipson (eds.), Inshore Fisheries Management, 207–220.

instruments differ between states reflecting the variations in political culture, institutional tradition and the importance and nature of the inshore sector. In some states, inshore fisheries are recognised as requiring special management structures or arrangements. Most, however, are governed by only partially devolved inshore management regimes and remain subject to regulations, often single species specific, applied also to the offshore sector. As indicated in the opening chapter - and demonstrated in the national reviews (Chapters 2 - 8) - the inshore management regime is distinguishable according to any one of a number of regulatory parameters including vessel length, target fishery and, only occasionally, geographical area (such as 6 or 12 mile limits or other territorial boundaries).

In part the framework for inshore fisheries management, and subsequent policy implications and impacts, is determined at the macro-institutional level. By virtue of the provisions in the European treaties, the European Union has exclusive competence in the management of sea fisheries. Member States, however, remain responsible for a number of activities and the principles of proportionality and subsidiarity have influenced the exercise of competence both at national and Community levels (Cudennec *et al.*, 1996; Morin, 1998). Thus, Council Regulation (EEC) No 3760/92 of 20 December 1992 establishing a Community system for fishing and aquaculture, states that 'the taking, implementing and monitoring of decisions should be done at the most appropriate level'. One area where this principle is applied is in the domain of access to Member States' territorial waters, authorising them, until 31 December 2002, to restrict access within a 12 nm zone. In certain defined areas of the 6-12 nm zone, access based on historic fishing rights is granted to vessels from other Member States. Member States are empowered to take conservation and management measures within waters under their jurisdiction insofar as these measures apply only to local stocks prosecuted by fishermen from the Member State concerned. In practice, therefore, Member States retain a degree of initiative in determining the details of management and the choice of institutional approach in relation to coastal fisheries. This is subject to the proviso that they conform with Community law together with the various measures (including total allowable catches, quotas, technical regulations, licensing, structural and marketing mechanisms *inter alia*) and overall objectives of the Common Fisheries Policy.

The macro-institutional influence over inshore fisheries is partially varied in the cases of the Baltic, which is affected by the International Baltic Sea Fisheries Commission, and in the Mediterranean, which comes under the auspices of the General Fisheries Commission for the Mediterranean as well as the European Union in the case of the Member States.

In most countries basic regulations providing for the foundations of fishing regimes have been established which, through the implementation of secondary legislation, allow for the setting up of specific measures that may vary in their field of application. Some of these measures are often initiated or implemented by organisational structures other than the central executive or Parliament and involve institutions to which some degree of authority has been devolved in order to contribute to the management of resources. In other instances these actors may play an informal or voluntary regulatory role. The same is true for the regulatory regime for inshore fisheries management which is executed using a combination of primary and secondary legislation. The delivery of inshore management can potentially embrace both public and private actors as well as both central and local levels of administration and regulation. In these terms four generic configurations can be envisaged for the delivery of inshore fisheries management (after Symes and Phillipson, 1996):

(i) *by central authority*: here the central state defines and implements all aspects of management; this may involve some form of consultation with local interests or organisations and the encouragement of informal or voluntary local management action;

(ii) *by decentralised authority*: this involves the transfer of management or administrative responsibilities from the central to the regional or local state - management responsibility remains within the democratically accountable political system; the framing of broad policy objectives rests with the central state and the decentralised authority aims to modify national instruments to suit local conditions;

(iii) *by delegated authority*: within limits set by the central state, certain responsibilities for management are delegated to local organisations outside of the democratically accountable system, such as agencies or private non-governmental organisations; co-management represents a specific example of delegated authority which involves the legally defined involvement of fishermen's organisations in an active management and decision making role;

(iv) *by autonomous authority*: here all management responsibilities originate with or are delegated to local user group organisations or communities; the basis for authority may be legally defined or based on traditional customary rights.

Systems of inshore management may, in practice, resemble individual or combined elements from the four generic configurations. A main dividing line is found between option (i) and options (ii), (iii) and (iv), essentially a division between centralised as opposed to locally driven regimes.

2. CENTRAL VERSUS LOCAL MANAGEMENT OF INSHORE FISHERIES

It is possible to identify several arguments for and against both central and local based approaches to the delivery of inshore fisheries management.

2.1 Local management regimes

A designated local management framework offers several hypothetical opportunities:

- the possibility of developing more sensitive and specific resource management, monitoring and enforcement approaches, based on an understanding of the local resource situation and fisheries politics, and allowing the development of appropriate contingencies for conflict resolution;

- the potential for more effective incorporation of local users in management activities, thus benefiting from user knowledge and improved legitimacy of the management system; the accommodation of different interest groups within the management process, and the encouragement of their willing compliance, may best be achieved at the local level where sectoral interests are more homogenous and social relations more developed;

- an improved sense of democratic accountability and alignment with the principle of subsidiarity in ensuring that decisions are taken as closely as possible to the citizen and at the most appropriate level (national, regional or local); locally based management helps to tackle the persistent image of a technocratic and divorced regulatory system (Vestergaard, 1993);

- the opportunity for reducing the regulatory burden of the central state and for redistributing the costs of management to other actors; and finally

- depending on the nature of the devolved management responsibilities, local management may provide an institutional basis for developing preferential approaches for locally dependent fishing communities; it may also provide a more effective means for the improved protection of sensitive inshore waters, habitats and juvenile fish stocks.

2.2 Centralised management regimes

Despite the seemingly strong case that can be made for local management, some arguments suggest that the most appropriate scale for the development and implementation of inshore fisheries management is the national level and undertaken by the central state. This allows for:

- less fragmentation of management responsibilities between organisations within the institutional framework and less requirement for harmonisation of management approach between disparate local organisations;

- the opportunity to take an overarching perspective of the management system and resource base; the central state maintains authority and control within the regulatory system;

- the perceived costs involved in establishing local systems; local organisations may be unwilling or unable to cope with the additional financial burdens, particularly with regard to enforcement;

- uncertainty over the organisational capacities of fishermen's organisations to receive delegated management responsibilities; in part this relates to the difficulties of generating collective action among sectoral and local user groups and the sometimes dissonant and fragmented nature of fishermen's organisations; and

- doubts over the democratic accountability of delegated management or co-management organisations and concern over the potential distributional implications of local management systems and the creation of privileged minorities; there is also a potential concern over signs of 'inverted tutelage', where a local administration assimilates the values and vested interests of groups placed under its control, which threatens the ability of the administration to take into account wider concerns (Bailly *et al.*, 1996).

The choice between a central or locally driven management regime for the delivery of inshore fisheries management is not necessarily a simple either/or scenario. For example, a centralised approach may also attempt to harness some of the benefits of local management through the use of regional enforcement structures, the enactment of local consultation or by the encouragement of local voluntary or informal management actions. In fact, it is unlikely that a state could define and execute management strategies without any recourse to or understanding of local circumstances, although the level of accrued benefits from informal or voluntary approaches would arguably be less than where local management is formally structured and legally defined. Similarly, the very nature of delegated or decentralised management configurations suggests a continued role for the central state which is able to maintain an overarching perspective or steering influence over the management system and resource base. Indeed, co-management implies a partnership based on trust and mutual respect between state and delegated authority. Hence, it is feasible that several of the benefits expected to emerge from centralised management may be in-built within local approaches.

3. INSHORE FISHERIES MANAGEMENT IN PRACTICE

The range of approaches adopted by the Member States under discussion is in practice very diverse. For example, the legislative basis for the measures and structures used in the delivery of inshore management does not reflect the 12 mile territorial waters. Instead individual states have devised their own particular arrangements to manage specific fisheries, sectors or geographical areas within or overlapping these limits. Furthermore, while some states rely primarily on a centralist approach, others have, to differing degrees, utilised delegated or decentralised structures. In these terms, a generalised allocation of national management approaches is posed in Figure 12.1.

From the figure it can be deduced that all inshore management approaches are subject to the influence of centralised direction, although to different extents. This is also the case where the system is characterised by delegation of responsibilities or decentralisation, where the central state maintains an overarching steering role. Furthermore, aspects of enforcement are commonly provided by regional divisions of the state inspectorate. It is also evident that there are no instances of autonomous management among any of the states under consideration.

The role of the central state is most pronounced in Denmark, Ireland and Scotland, where management is executed centrally by the respective

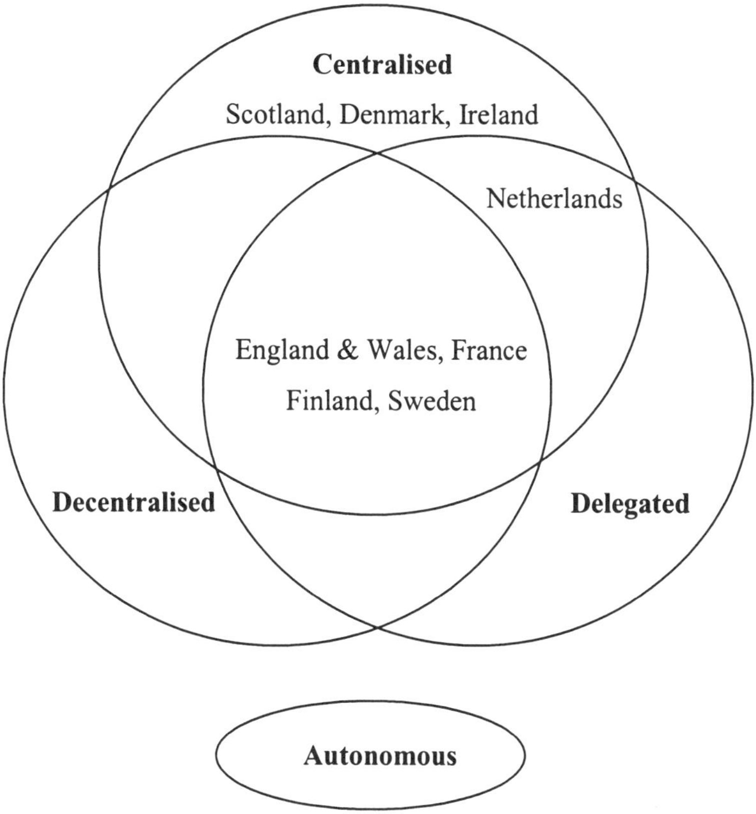

Figure 12.1 Regulatory Regimes for Inshore Fisheries Management

government departments and ministries (Danish Ministry of Agriculture and Fisheries, Department of the Marine and Natural Resources, Scottish Executive Rural Affairs Department), but in both Scotland and Ireland this situation is presently changing. In the case of Ireland, local inshore fisheries development committees are being monitored at three locations. If successful the arrangements may be replicated throughout Ireland, but on a voluntary rather than statutory basis. It is in Scotland, however, where the debate over central versus local management has been given a fresh impetus as a consequence of political devolution.

The newly established Scottish Executive has assumed responsibility for fisheries management within the 6 mile zone and has acted quickly to establish an inshore fisheries advisory committee involving representatives of the Executive, the fishing industry and the national nature conservation agency. It is too soon to forecast what changes in institutional arrangements or policy direction may follow. As outlined in Chapter 6, local fisheries management is at present confined to voluntary initiatives either in the form of area access management committees or more recently channelled through Regulating Orders which currently permit the introduction of licensing schemes and technical conservation measures but only in respect of shellfish. It seems unlikely that the earlier opposition to developing more formal structures for local inshore fisheries management, which comes largely from the offshore interests based in the north east of Scotland and has been endorsed by the former Scottish Office fisheries department, will be easily assuaged.

Devoid of formal devolved management structures, local involvement of user groups in all three countries is restricted to voluntary controls. Otherwise, the inshore sector's principal strategy for influencing management continues to depend on lobbying their own national fishermen's organisations to represent their interests in consultations with central government.

The Netherlands could arguably be placed in the same category as Ireland, Scotland and Denmark, as it is without a comprehensive inshore management framework and the sector is governed primarily by central policy guidance and control from the Ministry of Agriculture, Nature Management and Fisheries which applies individual pieces of legislation to specific fishing grounds within the Dutch territorial waters. The framework for inshore fisheries management is also steered at an international level within the context of the Trilateral Wadden Sea Conferences which form the basis for the incorporation of environmental objectives within the regulatory regime via the Dutch policy document on sea and coastal fisheries. However, delegation processes are also evident at both central and local levels. Centrally, this involves the National Fish Board, made up of representatives from all branches of the fishing sector, which has an important input into policy development. At a local level components of the inshore sector (essentially mussel and cockle groups) have organised themselves into producers' organisations and are active in the design and implementation of local management plans. Indeed, in terms of detailed management decisions, there are grounds for suggesting that the fishing plans drawn up by the producers' organisations represent a rare example in Europe of self-management, though they are, of course, subject to the approval of the central authorities. These plans have, not without some difficulties arising from the voluntary membership of the organisations, produced benefits in respect to both stock conservation and market planning (Chapter 5). They complement recent co-management initiatives within the Dutch industry in general (Langstraat, 1999).

Of the seven states under review the most developed structures for inshore fisheries management are found in England and Wales through the quasi-autonomous and uniquely constituted Sea Fisheries Committees and in France through the hierarchically structured *Comités des Pêches*. Both systems cut across the boundary between delegated and decentralised management configurations yet both are very different functionally and structurally (see Table 12.1).

Sea Fisheries Committees have a wide range of powers within their designated districts relating to the management, enforcement, monitoring and enhancement of fisheries. This locally sensitive regulatory remit includes the ability to propose and enforce byelaws and Several and Regulating Orders and is supported by a sea-going inspectorate which also monitors adherence to national and Community measures. This system is in sharp contrast to the centralised approach to the governance of inshore

Table 12.1 Functional and structural comparison of Sea Fisheries Committees and *Comités des Pêches*

	SFCS (England and Wales)	*Comités des Pêches* (France)
Membership	Unique structure of county council and Ministry appointee members. Appointees include fishermen, merchants, processors, sports fishing and environment interests (non-elective).	Multi interest representation (owners, crew, merchants, processors, producers' organisations, cooperatives). Representatives elected by registered seamen.
Mandatory membership	Constitution laid down in Statutory Instrument - membership not however mandatory.	Mandatory membership.
Hierarchy	Autonomous but nested within a national association.	Nested within regional and national committees.
Funding	Income from local government - not statutorily guaranteed.	Government funding and levy on landings.
Legal status	Sea Fisheries (Regulation) Act 1966.	Loi no. 91-411, 2 mai, 1991.
Geography	Based on 6 nm fishery districts. 12 SFCs based on local authority boundaries.	1 *Comité* for every marine quarter. 1 national *Comité*; 14 regional *Comités*; 39 local.
Regulatory functions	Regulation and policing of the inshore fishery as a whole (gear/vessel restrictions, licensing, minimum landing sizes, closed areas, stock enhancement activities etc.).	Contribute to management of inshore fisheries through the forwarding of advice and policy recommendations to regional committees and implementation of local regulations.
Welfare functions	None.	Promote social and economic conditions and training.
Other functions	Research and development.	Conflict resolution.

Source: after Symes *et al.*, 1996

fisheries presently adopted in Scotland. While both countries are relatively unique, in that they do define a 6 mile coastal band and provide specific legal provisions for the regulation of fisheries within this limit, the regulatory and institutional regimes clearly differ significantly.

The French approach is more complex; it is also less exclusive to inshore fisheries as no specific regulatory provision or institutional framework has been established to regulate the coastal fisheries, which are affected by the general legislation for French sea fisheries relating to licensing, total allowable catches and the application of the Common Fisheries Policy. Instead, provisions have been made to enable decision making on specific fisheries at the local level through a pyramidal and corporatist structure of organisation (Chapter 8; Thom, 1993). All participants in the fishing sector are required to register as members of the local *comités locaux des pêches*, which nest within regional and national levels of organisation. In addition to representing the interests of their constituents they contribute to the management of the resource base in collaboration with the regional and local Affaires Maritimes, a decentralised level of administration. Proposals on resource management and licensing arrangements may be forwarded to the local office of the Affaires Maritimes or alternatively to the regional or national *comités*; as such it is a 'vertically oriented' system in emphasis. Local *comités* are partly responsible for the implementation of regulations relating to the control of fishing activities framed at regional and national level, though powers of policing and enforcement remain in the hands of the local and regional administration. While the administration maintains control over whether to sanction the recommendations of the *comités*, some observers argue that the actual level of administrative authority is tempered by the difficulty they would experience in rejecting a decision taken by the *comités* (Curtil, 1993; Bailly *et al.*, 1996).

The situation on the French Mediterranean coast is further complicated by the enduring system of traditional fishermen's organisations or *prud'homies* which, in addition to representing their members in negotiations, also have a role in fisheries management at the community level (Dufour, 1996; see also Chapter 8). Like the *cofradias* of Spain, they play an instrumental role in defining access arrangements, allocating fishing rights, specifying gear methods, solving local disputes and administering local justice, making 'strong imprints on the modes of management of fishing grounds and on the social life of the fishing communities' (Dufour, 1996: p. 110). Though their remit is increasingly constrained by the central state and the *comités,* the *prud'homies* have remained semi-autonomous and feature as an integral component of the local industry and community.

Attempts to classify the general approaches to inshore fisheries management in Finland and Sweden are uniquely complicated given the separation of management responsibility between state controlled waters and coastal waters under private ownership (Chapters 2 and 3); here the management regime for freshwater and coastal fisheries is determined by the ownership patterns of the waters. In both states, though to a greater extent in Finland, there are elements of both decentralised and delegated management. In Finland responsibility for management and data collection within private coastal waters is delegated to fisheries regions, each under the authority of a fisheries council comprising statutory shareholders associations (for areas held in common by a village), individual private owners, commercial and recreational fishing interest organisations and occasional public authorities. They represent an intermediate level of administration with public duties, but are not a decentralised branch of government (Sipponen, 1999) and remain subject to the management parameters set by national legislation. Outside Finnish coastal waters managerial tasks are performed centrally by the Ministry of Agriculture and Forestry or decentralised to the governmental regional fisheries administration (Employment and Economic Development Centres) which also coordinate the activities of fisheries regions. In practice the coastal fisheries councils have not taken a proactive approach to fisheries management when compared to their inland lake counterparts. The reasons are several and include the role played by the European Union or International Baltic Sea Fisheries Commission in determining catch quotas for certain species, the difficulty of managing migratory resources on a regional basis, a reluctance on behalf of actors to use the powers available to them and slow political approval of the fishing region system (Sipponen, 1999). To some extent comparable arrangements exist in Sweden though the regions involved are smaller and less developed. Management in coastal waters is the responsibility of private owners (individuals, communities or associations) working within legislation set by the National Board of Fisheries which is responsible for fisheries management in public waters. Limited administrative responsibilities are also decentralised to county administrations (see Chapters 3 and 16).

4. ANALYSIS AND CONCLUSIONS

As a whole the countries analysed in this volume display considerable diversity in the level of local structure that they afford to inshore fisheries management, in the degree of central state involvement and in the patterns of interaction between state authorities and the fishing sector. This distinctiveness of regime is likely to reflect a combination of elements including the particular characteristics of the fishing sectors, the traditions of state intervention and user participation, the sectoral imperatives intrinsic to the country concerned, the level and history of organisational development and the perceived balance of costs and benefits arising from local and central approaches. The outcome of such a 'cost-benefit' analysis may reflect the internal politics of the fishing industry and the interests of particular sectors, or the perceived financial costs of local management. It might also be based upon a groundswell of opposition to proliferating 'corporate led' local management approaches, which are seen to signify a democratic deficit by privileging certain groups above others in terms of access to the policy community or fishing grounds. Furthermore, the political will may not be sufficiently developed in some states to support the development of coherent approaches to inshore fisheries management.

The extent to which a state has a well defined or distinctive inshore sector, and the actual levels of fishing activity involved, may also be important in justifying the development of local regimes. In some instances the difficulties implicit in regulating inshore fisheries - which may be characterised by low catches, large numbers of units, poorly developed statistics, and undefined activity rates arising from the part time, seasonal or pluriactive nature of participation - may argue in favour of a centralised approach.

The state of local organisational development may not always be considered sufficiently developed or secure to allow local approaches to management based on the involvement of local fishermen's organisations. In some states such organisations may appear fragile or lacking in organisational capacity, possibly reflecting a lack of organisation among often individualistic and geographically dispersed enterprises, or an incomplete membership coverage. Other countries may be characterised by a strong centralist tradition in organisational development, whereby local fishermen organise themselves within a national representative organisation rather than local or regional bodies, and here there may be little sense of a need for local structures. In some cases local approaches to management may be best developed through the mobilisation of local interests by established decentralised authorities or through the regional division of central representative organisations. However, the political influence exerted by the more powerful offshore fishing interests both within their own organisations and on fisheries departments may be sufficient to block moves to establish local management structures.

The occasions where inshore fisheries are subject to local management also appear to vary in terms of the scope of the delegated or decentralised management remit. In most cases local management appears to involve industry management rather than resource management. While the division between the two is not always clear cut and may reflect an assessment of the overall orientation and breadth of activity rather than any particular division of management tasks, it is feasible to generalise different approaches on this basis. Industry or sector management refers primarily to the organisation of access and fishing rights (through licensing or boundary delimitation) and the settling of gear disputes. Only in a few cases is the scope of local management extended to include a wider resource management remit. The Sea Fisheries Committees of England and Wales may represent the most holistic local approach to inshore management embracing - albeit subject to certain financial and procedural limitations - regulatory powers, stock enhancement, research and monitoring activities, and policing and enforcement responsibilities. They also play an active role in marrying fisheries regulation with concerns for the marine environment.

There are several common regulatory challenges facing local arrangements in the future. They involve, for example, clarification of the future of the 6 and 12 mile access derogations in 2002 which represents a source of significant uncertainty within the inshore sector and a potential barrier to long term planning. However, as the European Commission has noted, 'it seems unlikely, given the current state of affairs, that there will be a desire to modify this aspect of the *acquis communautaire*. So while agreement on a new Commission proposal for 12 mile limits will be required in order for those limits to be maintained, our current assessment is such that a proposal would be desirable and virtually certain to obtain the necessary qualified majority in favour in the council' (*Fishing News*, 29 May 1998). Renewal of the derogation may signify a step towards a strengthening of preferential access for local vessels adjacent to coastal waters or even engender a clearer separation of management structures and regimes between inshore and offshore fisheries and a shift in the balance of empowerment between the centre and locality.

It is most probable, however, that the key challenge facing local management will relate to the question of management scope and more particularly the positioning and level of involvement of local institutions with regard to new areas of management activity. Central to this will be moves towards integrated management, involving the harnessing of fisheries and environmental management objectives, and coastal zone management, involving the coordinated planning of multiple use activities along the coast (see Chapters 14 - 16). In fact, perhaps one of the strongest arguments for local as opposed to central approaches is the potential that such regimes might offer for securing effective approaches to an integrated system for coastal zone management. The question arises as to whether these new responsibilities can be delivered through existing structures. If so, then what structural adaptations do they need to implement to embrace the extended remit? If not, or if local structures are not presently available, then how and in what form should integrated management be attempted?

5. REFERENCES

Bailly, D., Boncoeur, J., Guyader, O. and Le Floc'h, P. (1996) *Economie et Droit des Ressources Naturelles Renouvelables de la Mer, Aspects Theoriques et Application a la Zone Cotiere de la Manche Occidentale Francaise*, Volume I, Etude Economique, CEDEM, Brest, December 1996.

Cudennec, A., Curtil, O., Labrot, V. and Prat, J.-.L. (1996) *Economie et Droit des Ressources Naturelles Renouvelables de la Mer, Aspects Theoriques et Application a la Zone Cotiere de la Manche Occidentale Francaise*, Volume II, Etude Economique, CEDEM, Brest, December 1996.

Curtil, O. (1993) *Environnement Juridique de la Filiere d'Elevage de la Coquille Saint-Jacques*, IFREMER/Université de Bretagne Occidentale, Brest.

Dufour, A-H. (1996) Management and practice in the small-scale inshore fisheries of the French Mediterranean, in D. Symes and K. Crean (eds.), *Fisheries Management in Crisis*, Blackwell Science, Oxford, pp. 110-115.

Langstraat, D. (1999) The Dutch co-management system for sea fisheries, in D. Symes (ed.), *Alternative Management Systems for Fisheries*, Blackwell Science, Oxford, pp. 73-78.

Morin, M. (1998) Legal elements of fisheries management in the western Mediterranean. Paper presented at the European Social Science Fisheries Network Workshop on *Southern Waters: Management Issues and Practice*, Syros, 14-16 May 1998.

Sipponen, M. (1999) Fisheries regions: an organisational structure for fisheries management in Finland, in D. Symes (ed.), *Alternative Management Systems for Fisheries*, Blackwell Science, Oxford, pp. 51-59.

Symes, D. and Phillipson, J. (1996) The imperative of institutional reform: alternative models and the UK fishing industry, in *Proceedings of the VIIth Annual Conference of European Association of Fisheries Economists,* CEMARE, Portsmouth, pp. 229-243.

Symes, D., Crean, K. and Phillipson J. (1996) *Devolved and Regional Management Systems for Fisheries: Final Report*, AIR-CT93-1392, School of Geography and Earth Resources, Hull.

Thom, M. (1993) The governance of a common in the European Community. The Common Fisheries Policy, PhD thesis, University of Strathclyde.

Vestergaard, T. (1993) Catch regulation and Danish fisheries culture, *North Atlantic Studies*, **3** (2), 25-31.

PART III: MANAGEMENT ISSUES

Chapter 13

THE ROLE OF AQUACULTURE IN INSHORE FISHERIES

Nathalie Steins and Eva Roth

1. INTRODUCTION

Aquaculture is an old established system of food production. The Romans were reputedly farming shellfish in the Mediterranean coastal lagoons some two thousand years ago; freshwater fish husbandry was developed on the monastic estates of western Europe during the medieval period. But it is only with the rapid expansion of marine aquaculture in the last quarter of a century that it has come to be recognised both as a significant contributor to global food supplies and as a potential problem for the aquatic environment.

Over the past two decades, the aquaculture industry has witnessed tremendous growth. Between 1984 and 1998, the world output of farmed shellfish and finfish quadrupled to 33 million tonnes. In 1998 aquaculture produce accounted for 26% of the world's total fish supply (FAO, 2000). In the European Union (EU), output in 1997 totalled 1.2 million tonnes and European aquaculture was responsible for the employment of 47,000 workers (European Commission, 1999). Significantly the expansion in output within Europe was occurring at a time of declining production from the region's capture fisheries.

Traditionally, global aquaculture production has been dominated by Asia: China, India and Japan produced 57%, 9% and 4% of world output in 1994, respectively. Until the late 1970s Asian production was largely intended for the domestic markets. From the 1980s onwards there has been increasing interest in the production of farmed seafood for the American and European markets. At the same time, aquaculture production in the United States and Canada and in the north west of Europe (Norway, Scotland, Ireland) has grown rapidly (Weber, 1997). The bulk of Europe's aquaculture production

D. Symes and J. Phillipson (eds.), Inshore Fisheries Management, 223–238.

occurs in Norway and the EU and is dominated so far by the production of high value species. Molluscan aquaculture, accounting for some 620,000 tonnes in 1995, is based largely on blue mussels (*Mytilus edulis*), Mediterranean mussels (*M. galloprovincialis*), Pacific oysters (*Crassostrea gigas*) and clams (*Tapes* spp). But the most rapidly growing aquaculture production in Europe concerns the salmonids (614,000 tonnes in 1995) - the Atlantic salmon (*Salmo salar*) and rainbow trout (*Onchorynchus mykiss*) - farmed principally in north west Europe. Other finfish species include the sea bream (*Sparus auratus*) and sea bass (*Dicentrarchus labrax*) grown principally in southern Europe (Ross and Roed, 2000). Of potentially much greater long term significance for Europe's capture fisheries is the farming and ranching of some of the key target species, including cod, haddock and halibut, presently in the research and development phase.

Aquaculture development has been encouraged world wide for a number of reasons, the most important being employment generation in peripheral coastal communities, the growing need and demand for domestic food supplies, the expansion of export earnings, the reduction in pressure on wildfish stocks, and the enhancement of recreational and commercial fisheries through the restocking of natural habitats. In the EU, the cost-benefit of aquaculture development must be set within the context of the Common Fisheries Policy and related to aquaculture's role in supplementing production from depleting fish stocks, free from quota restrictions, and offsetting the widening trade deficit in fish and fish products. Aquaculture is identified as a specific target for support in the EU through the Financial Instrument for Fisheries Guidance (FIFG).

Despite some obvious economic benefits, aquaculture is the subject of an intensifying debate. In the first place, aquaculture is associated with a number of negative externalities mainly impacting on the marine environment and its ecosystems (see Bailly and Paquotte, 1996; Black *et al.*, 1997; Commission of the European Communities, 1995; Goldburg and Triplett, 1997; ICES Mariculture Committee, 1998; Ross and Roed, 2000; Weber, 1997) and the use of marine space by other users (Bailly and Paquotte, 1996; Commission of the European Communities, 1995; Steins, 1998). Secondly, the future prospects for aquaculture's growth may not be quite as rosy as they have been painted in the past; according to one report on the EU (Stirling Aquaculture Ltd., 1996) there is scope for major development only in four of its Member States.

The prime objective for the development of culture fisheries is to provide society with the economic and social benefits of fish farming under sustainable conditions. To counteract certain economic, social and environmental externalities, management frameworks are gradually being developed and implemented. The present chapter explores the interactions between capture and culture fisheries in inshore waters, outlining both the advantages and disadvantages and exploring future strategies within a European context.

2. AQUACULTURE AND ITS ADVANTAGES FOR INSHORE CAPTURE FISHERIES

Aquaculture development is often seen as a direct response to problems facing the capture fisheries. Fish stocks are depleting on a global scale as the result of natural, human and institutional factors. The CFP, embracing structural, marketing, conservation and external dimensions, has striven to provide a strategy to address the problem. Sometimes regarded as a political success (Holden, 1994), the CFP must counter strong market forces which conspire to create problems of overcapacity and overfishing. Economic pressures have helped to shape the modern fishing industry. Fish, along with other food for human consumption, has declined in relative value and a rise in labour productivity is necessary to counteract falling prices and safeguard income for consumption and investment in fishing households. This has resulted in higher capital-labour ratios in the fishing industry and lower employment in the harvesting sector. Combined with depleted stocks, excess fishing capacity and poor performance in certain segments of the fleet, it engenders a further slackening of fishing activity and still more job losses. The impacts are most keenly felt in some of the remoter fisheries dependent regions generally characterised by peripherality, poor infrastructure and a strong dependence on the inshore fishing sector (Symes, 2000). It is precisely in this kind of context that aquaculture is seen as a way of generating alternative employment and income earning opportunities and reducing the pressure on relatively scarce wildfish stocks.

Public perception of aquaculture, and particularly the newer element of finfish farming, has been influenced by negative assessments of its environmental impact dating mainly from studies conducted in the 1980s since when the industry has markedly improved its practices. The percentage of waste feed discharged into the marine environment and the use of antibiotics to control disease has decreased by several orders of magnitude. Recent monitoring programmes in Maine and Washington, USA, and elsewhere have documented increasing biomass and abundance of benthic invertebrates and fish around the salmon cages. Aquaculture can interact positively with other marine users such as the tourism industry, through the improvement of recreational fisheries. Table 13.1 summarises the positive effects of aquaculture on the natural and socio-economic environments; some of these are of particular importance to the inshore capture sector.

Table 13.1 Positive interactions among aquaculture and other uses of the coastal zone

Benefits	*Type of Mariculture*					
	Freshwater hatcheries	*Saltwater hatcheries*	*Nearshore fish farms (cages)*	*Nearshore shellfish suspension culture*	*Nearshore shellfish bottom culture*	*Offshore finfish and shellfish culture*
Employment*	✓	✓	✓	✓	✓	✓
Tourism revenue			✓	✓		✓
Recreational fishing	✓	✓	✓		✓	✓
Water quality	✓	✓	✓	✓	✓	✓
Education	✓	✓	✓	✓	✓	✓
Stock enhancement	✓	✓		✓		

*Employment includes rural development, transportation, packaging and appropriate income multipliers
Source: ICES Mariculture Committee, 1998

2.1 Employment and income opportunities

In many coastal regions, employment opportunities can be scarce. In the more remote areas of Europe like western Ireland and north west Scotland, climatic and physiographic factors limit the development of agriculture and tourism while poor transport and marketing infrastructures constrain the growth in manufacturing and service industries. In such areas fishing has traditionally been an important means of income generation and the continuing reduction in the level of activity has had a severe impact on some fishing communities. The development of culture fisheries can make a significant contribution to communities facing these challenges. As outlined in Chapter 7, the structural problems of the Irish fishing fleet combined with marketing problems, are resulting in increasing unemployment rates among fishermen while those engaged in related shore based activities are also experiencing difficulty in finding jobs. With the assistance of the EU's PESCA initiative, 52 shellfish cultivation projects involving individuals or cooperatives have been established in Ireland. For the majority of the fishermen involved, shellfish farming is a means of supplementing their incomes from traditional fisheries, though some have abandoned the capture sector altogether. The cultivation activities focus either on the complete production cycle as in the case of mussels or on restocking the depleted wild shellfish stocks as with scallops and native oysters. Five training programmes in shellfish cultivation have been initiated in Ireland.

2.2 The reduction of pressures on wildfish stocks

Development of culture and capture fisheries is not mutually exclusive. As the natural reservoirs for the exploitation of wild stocks become exhausted, aquaculture is increasingly seen as a way to enhance production from finfish and shellfish populations. Shellfish cultivation techniques are increasingly used to rehabilitate existing wild fisheries; in Castlemaine Harbour in Ireland, for example, natural scallop beds have been successfully restocked using cultivated scallops. Similar projects are taking place in other shellfisheries using hatchery cultivated seed. In many recreational salmon and trout fisheries in Europe and North America, native broodstock is being reared in hatcheries for release into the rivers. Ranching is gaining in popularity as a means of improving recreational fishing opportunities while keeping the impact on wild runs as low as possible. In the case of the native American tribes fishing the Columbia River in the USA, where salmon runs had declined due to dam construction, urbanisation and deteriorating conditions in their oceanic phase resulting in the listing of salmon as an endangered species, hatcheries have enabled the tribal populations to continue their salmon fishery which forms a crucial part of their cultural identity. The link between aquaculture and capture fisheries can, of course, also work the other way round: wild caught fry is an important input into the aquaculture production of eels and shrimps.

2.3 Use of existing marine resources

Aquaculture production relies heavily on the natural marine environment and on existing natural resources. Shellfish production often requires no external inputs other than manmade production structures (longlines, rafts or trestles) and is thus relatively easily accessible to inshore fishermen. For finfish farming other external inputs, such as feed and antibiotics, are needed to maintain the production cycle. Unlike agriculture, however, neither form of aquaculture requires significant modifications of the natural environment; the basic potentials of the natural resource base can be utilised and further expanded. Nevertheless, as will become clear in the discussion of the potential problems of culture fisheries for inshore capture fisheries, alternative uses for inshore waters, limits to the carrying capacity of inshore environments and the decreasing availability of suitable new sites now form constraints on further aquaculture development in many areas.

2.4 Well defined property rights

Aquaculture production usually takes place in clearly demarcated licensed areas, that is on sites where property rights adhere. By denoting a resource as 'property', a reservoir or flow of individual benefits is identified. Property rights are social institutions which have evolved as a means of enforcing claims to the benefit stream, and by attaching rights to property the intention of enforcing duties on would be users to observe restricted or prohibited access to the resource is clearly demonstrated (Edwards and Steins, 1998). In a study of the role of property rights in the fishing industry, based on the assumption that rights are considered 'property rights' when they are exclusive, LEI-DLO *et al.* (1997) identified three types of property right: state property; private property; and common property. 'Open access' can be interpreted as a situation of non-property.

The existence of property rights in the aquaculture industry is associated with three advantages. First, it is clear where a site is located and who owns it, making control and enforcement easier for agencies like the national fisheries departments. Second, the allocation of property rights - either private or common - is believed by some to contribute to better stewardship of the resources and greater compliance with regulations. Third, individuals who do not hold property rights can be excluded, thus removing the problem of the freerider. Property rights do not of themselves guarantee sustainable management: as Bromley (1991: p. 2) has observed: 'property is not an object such as land but rather a right or group of rights to a property stream that is only as secure as the duty of all others to respect that stream'. Furthermore, as in the case of marine fisheries, property rights regimes in aquaculture were usually not explicitly determined before the management schemes were implemented. This has resulted in unintended effects in terms of the structural and economic development of aquaculture. Property rights to aquaculture sites tend to be granted for limited periods of time with no guarantee of renewal, which means that investment costs may have to be spread over a shorter time span than is desirable from a management perspective. The poor development of management regimes for aquaculture development has also resulted in social externalities such as conflicts between aquaculturalists and fishermen (Steins, 1998).

3. AQUACULTURE AND THE DISADVANTAGES FOR INSHORE CAPTURE FISHERIES

While there may be several benefits to be derived from aquaculture, its development is also associated with certain problems, mainly concerned with the emergence of conflicts of interest between traditional and new forms of economic activity in coastal waters - in particular the reduced access for traditional fishing activities - and the negative impacts of aquaculture production on the natural environment, which stem largely from finfish aquaculture. Some of these issues, especially those relating to multi-use conflicts, are also considered in Chapter 15.

3.1 Falling market prices for capture and culture fisheries

There can be little doubt that the dramatic increase in the supplies of farmed fish to the world's markets has had a negative effect on the first hand sales of fish at port markets. In Europe the increasing availability of farmed salmon has resulted in a substantial fall in the market price for wild salmon and probably impacted quite significantly on the market value of many other popular whitefish species. In some instances fishermen may attempt to compensate for falling prices by increasing fishing effort, at least in the short term. In the long run, however, falling prices are more likely to result in a redeployment of resources and a reduction in fishing effort on the species most directly affected. In Ireland where prices for wild salmon have shown themselves to be inversely proportional to the growth in farmed salmon production (Fingleton and McCann, 1996), salmon netting has become more of a recreational activity or simply a means of maintaining a family tradition, as the economic benefits of the enterprise no longer outweigh the costs of production. Shifting to other fishcrics may not be feasible because of licensing restrictions and a lack of capital for reinvestment in the new fishery. In such circumstances inshore fishermen may well be marginalised and eventually driven out of business as the result of aquaculture's seemingly incessant growth. On the other hand, new jobs are being created in the aquaculture sector and consumers are reaping the benefits of low prices for what was once a luxury food. In Europe, certainly, salmon has become increasingly accessible in terms of price and its consumption has been increasing.

The increased production of shellfish on a global scale has also resulted in a fall in the market price for shellfish products. Over the past five years, the price for flat oysters, traditionally regarded as an exclusive product, has dropped by over 50%. Lower production costs have led to lower market prices and higher volumes being sold on the market. In addition, the cheaper and more easily produced Pacific oyster has successfully competed in the market for oysters.

3.2 Accessibility

Although aquaculture is often promoted as a means of economic development in remoter coastal areas, it is not always economically accessible to inshore fishermen and their communities. In particular, finfish farming is subject to high capital investment and demands specialised knowledge. Not surprisingly, therefore, most of the larger salmon farming enterprises in western Europe are either directly owned by, or at least affiliated to, large multi-national corporations in the food and/or chemical industries. However, shellfish aquaculture is more easily accessible to inshore fishermen at least in terms of the required capital investment. Grant aid is available in most countries. EU grants supported the diversification of traditional fisheries through the PESCA initiative (1994-1999) and funding is still available through FIFG and the revised Objective 1 and 2 frameworks. However, the lead time between initial investment and the first harvest, the vulnerability of shellfish to disease and the absence of a sound marketing infrastructure are major constraints to small scale entrepreneurs becoming involved in this type of activity. As a consequence, new shellfish projects are quite often initiated through the concerted action of several fishermen in the formation of a cooperative or small company, in order to pool resources, secure financial backing and reduce economic risks.

3.3 Scarcity of suitable production sites

Aquaculture places high demands on the sites chosen for production: they have to be sheltered, deep (especially in the case of intensive mussel and salmon cultivation) and should be located in good quality water of moderate temperature. The expansion of aquaculture has resulted in a scarcity of suitable sites. One way of overcoming this problem is to intensify production on existing sites but this strategy may be frustrated by two factors: first, there is a limit to the amount of shellfish cultivation that any estuary or bay can sustain; and secondly, in some countries, salmon production has already reached maximum levels of production as far as the environment is concerned. Another solution is to move aquaculture operations offshore, but here technical considerations and the financial outlays involved pose serious constraints for the implementation of this approach, at least in the immediate future. The scarcity of sites also has repercussions for the increasing tendency to incorporate fallowing sites in salmon farming in order to prevent the build-up and spread of disease. The lack of sites means that many farms may have to reduce their production levels in order to facilitate the rotation of production areas.

3.4 Environmental and ecological externalities

A final problem for inshore capture fisheries concerns the environmental externalities of aquaculture. Aquaculture activities are considered a potential source of eutrophication, resulting from faecal waste and unconsumed fish food (Commission of the European Communities, 1995; Goldburg and Triplett, 1997; Weber, 1997). If phytoplankton production exceeds the ecosystem's assimilative capacity, so-called algal blooms may result. The occurrence of 'red tides' - toxic blooms which can kill a wide range of aquatic organisms and accumulate in certain shellfish causing health risks for consumers - is not directly linked to excessive nutrient release (Black *et al.*, 1997). Nonetheless algal blooms can prove disastrous for both aquaculture producers and for shellfish fishermen; they can affect the benthic environment which provides the natural shellfish beds or nursery grounds for capture fisheries. Chemicals used to control disease, parasites or weeds in finfish production tend to disperse in the waters surrounding the production sites and can accumulate in both the sediments and the benthic organisms; copper based treatments have, for example, been linked to natural shellfish mortalities (Goldburg and Triplett, 1997; Weber, 1997). Outside Europe in areas where standards of monitoring, control and enforcement are poor, there is a much greater risk of deliberate habitat destruction - as with the reclamation of mangrove swamps to allow for the development of intensive shrimp farming - with potentially very serious consequences for the environment and its sustainable exploitation (Bailly and Paquotte, 1996).

4. INTERACTIONS BETWEEN CAPTURE AND CULTURE FISHERIES IN INSHORE WATERS

From the foregoing analysis it will be evident that the benefits of aquaculture do not fall neatly into one category and the drawbacks into another; the advantages and disadvantages are commonly reciprocals of the same situation. Dynamic interactions between culture and capture fisheries occur on several levels - ecological, social, economic, cultural and political - yet aquaculture is often considered in isolation from traditional capture fisheries. This projection of aquaculture as a separate activity is fed by a number of factors.

First, it is difficult to visualise aquaculture as 'fishing'. The production cycle is to a large extent controlled by the producer. The harvesting of farmed finfish does not remotely resemble the activity of fishing; nor is it subject to 'management' in the shape of output or input limitations or technical conservation measures as is normally the case in capture fisheries. Most regulations applied to aquaculture - once licences have been granted - refer to consumer health and environmental quality. Many aquaculturalists see themselves as 'cultivators' and, indeed, may resent the term 'fishermen'. Secondly, due to the levels of investment, knowledge and technology involved, finfish aquaculture tends to be the domain of large multi-national companies with little or no relation to the fishing industry. Ownership patterns differ from those in the inshore capture sector: in contrast to the dominance of small family firms (Chapter 9) and simple commodity production (Chapter 11), the aquaculture sector is characterised either by cooperative arrangements between self-employed individual producers, as in the shellfish sector, or capitalist firms involving wage labour in the finfish sector. Thirdly, the management issues commonly associated with aquaculture differ substantially from those in capture fisheries and demand a wholly different approach. Indeed, the management issues are more likely to be comparable to those of the agriculture sector.

But to consider aquaculture as a wholly discrete activity does not do justice to the dynamic nature of the inshore fisheries. The use of cultivation techniques to enhance production in inshore shellfisheries is historically well established. In the Netherlands, for example, mussel seed was transplanted to special cultivation parcels as early as the mid-nineteenth century, a practice on which the Dutch mussel industry still depends today (van Ginkel, 1991). Further, shellfish aquaculture has in many cases become complementary to inshore fishing and is embedded in the everyday realities of fishermen and their families. In some fishing communities in the remote areas of western Ireland and north west Scotland, aquaculture has largely taken over from traditional fishing activities. Without aquaculture it is questionable whether such communities could survive in the long term since fishing opportunities are declining and alternative employment is scarce especially outside the tourist season. In some inshore shellfisheries aquaculture techniques have become indispensable to the rehabilitation of traditional wild fisheries that have suffered from overexploitation. Cultivation techniques have also assisted the upgrading of recreational fisheries, most notably salmon and trout, which provide important sources of income generation in the tourism sector. Finally, the adoption of aquaculture techniques to improve the productivity and market quality of shellfish has helped to expand the consumer markets to the benefit of culture and capture fisheries alike.

Capture and culture fisheries are not mutually exclusive but complementary, though to date the interdependence is much more strongly evident in the shellfish rather than the finfish sector. It would be difficult to think of the inshore waters of Norway, Scotland, Ireland, France and the Netherlands today without the presence of aquaculture in some form. 'Farming the sea' has become part of everyday fishing practice for many inshore fishermen in Europe and elsewhere. Understanding the challenges that fishing communities currently face - and the responses of those involved - is not possible without first considering the dynamic relationships between capture and culture fisheries. Yet not infrequently these two sectors are handled by different government departments and only rarely are integrated strategies for their development and management put forward.

5. AQUACULTURE MANAGEMENT

In reality there is little common ground between the management needs of capture and culture fisheries. In the former case management has come to mean intervention by central or regional authorities to reduce harvesting capacity, restrain fishing effort, limit fishing gears and generally restrict production in order to protect the fish stocks from the risks of overexploitation. By contrast, until very recently, governments have actively encouraged the expansion of aquaculture and, subject to the availability of sites, have been generally willing to grant licence applications for the setting up of new production units or the extension of existing ones. Decision making over the granting of licences has not always been fully transparent: few countries have adopted a zonal approach distinguishing between areas where aquaculture development would be welcomed or actively discouraged; and local planning enquiries have only recently been based on a thorough socio-economic and environmental cost-benefit appraisal. Van der Schans (1996), using a Habermas perspective, explains the continuing controversy over salmon farming in Scotland as resulting from a lack of communicative rationality in the planning process. Only rarely have institutions intervened in an attempt to control the growth of production in finfish farming either to try and stabilise the markets, as in Norway at the start of the 1990s, or because of fears over the health of the aquatic environment (see Chapter 4).

According to van der Schans (1999: p. 117) the development of marine finfish farming

> 'cannot be looked upon as an isolated activity to be dealt with by a single regulatory agency. Rather it requires co-ordinated action of public authorities and private parties to cope with a diversity of interests in relation to a complex and highly dynamic marine ecosystem'.

The problem is to construct an appropriate organisational framework for such a complex and technically demanding task and to identify the lead agencies. The challenge of management is to realise the socio-economic benefits while reducing the social and environmental costs associated with aquaculture. In practice, this has meant focusing essentially on the environmental externalities though, as the debate over aquaculture continues to unfold, these externalities seem to increase rather than diminish.

Nevertheless, significant efforts are being made to reduce the negative side effects of aquaculture. Attention is increasingly being given to the choice of production sites, taking into account water circulation, the establishment of minimum distances between fish farms and the provision of fallowing areas. Single bay management strategies are becoming common practice. The use of chemicals has been greatly reduced; Norwegian producers, for example, had reduced the use of anti-microbials by 90% between 1987 and 1994 (Black *et al.*, 1997). But management can be pushed much further. There is scope for the adoption of an ecosystem approach, integrating the objectives of sustainable fisheries development and marine wildlife conservation (see Chapter 14) and, even more broadly of locating management decisions concerning aquacultural development within a framework of integrated coastal zone management, wherein participation by all relevant user groups promotes a sense of stewardship of the inshore marine environment.

Approaching the issue from very different perspectives, two recent reports have highlighted the need to extend the management process and incorporate stakeholders more fully within decision making. Both reports were concerned to establish a more robust framework for an integrated approach to the sustainable development of aquatic resources. Reviewing the proceedings of a Europe wide symposium on inland fisheries and aquaculture, the European Inland Fisheries Advisory Commission (1998) identified a number of key areas for action, including:

- the need for conventional aquaculture systems to evolve and adapt to changing social, economic and environmental conditions;
- the need to define a regulatory framework in which aquaculture is accepted and legally recognised as a legitimate user of water;
- the prioritisation in R & D programmes of water efficient fish production systems to allow for a more sustainable use of water resources; and
- the integration of scientific research, environmental appraisal and monitoring, improved communications between users and the involvement of local stakeholders in the policy and planning processes to promote a sustainable use of water resources by aquaculture at an optimal level of exploitation, acceptable to all users, whilst maintaining the potential to meet the direct requirements of rural and coastal communities as well as the needs of future generations.

In the same year, the ICES Mariculture Committee (1998) published its report on the environmental interactions of marine aquaculture. Among its recommendations were:

- the greater use of geographical information systems (GIS) as a decision support system in coastal zone management;
- a socio-economic assessment of the role of aquaculture in order to establish the relative importance of the sector for coastal communities (involving the development of green input-output tools);
- recognition of the value of stakeholder involvement in the development of integrated coastal management strategies, while appreciating the weaknesses implicit in the participatory process; and

- an increasing awareness among local authorities of the need to set aside traditional land use planning approaches when taking responsibility for integrated coastal management.

Although the dynamic nature of interactions between aquaculture and the wider socio-economic and ecological systems is increasingly recognised, the precise role of aquaculture in coastal fishing communities has been somewhat underexposed. Research has focused primarily on the maximisation of production efficiency and the reduction of environmental externalities. The role of aquaculture in sustaining the livelihoods of fishing communities through job creation and stock enhancement is frequently referred to but the allusions are based perhaps more on aspiration than on scientific evidence. There is a need for economic and social research to elaborate the interactive processes between culture and capture fisheries, focusing not only on areas of conflict and the means of resolution but also on the economic and social consequences of developing culture fisheries in inshore fishing communities.

6. CONCLUSION

Aquaculture today forms an integral part of the economic life of many coastal communities, as it has done for centuries in some parts of Europe. Only quite recently, however, have modern forms of intensive finfish farming experienced accelerated expansion aided by technical innovation and actively stimulated by governments and development agencies. The rapid development of aquaculture in the modern era is associated with a number of perceived advantages for coastal communities caught up in the spiral of decline affecting many of the traditional capture fisheries: these include the creation of direct and indirect employment, the protection or enhancement of certain commercial and recreational fisheries and the enhancement of product quality. But its development may also bring adverse effects for traditional fishing communities, including environmental degradation and concerns for the sustainability of marine biodiversity, and a shift in the seafood supply curve spelling lower market prices for both culture and capture species.

The interactive nature of capture and culture fisheries is now acknowledged and efforts are being made to balance aquaculture with other interests in inshore waters. The strategies adopted by inshore fishermen are inevitably affected by the opportunities and constraints inherent in the development of culture fisheries. It is surprising therefore that the net benefits of aquaculture for inshore fishing communities are not better documented. The role that aquaculture can play in the socio-economic development of fisheries dependent regions is still based mainly on anecdotal evidence and assumptions. Only quite recently has the focus of attention shifted from purely biological and economic matters to social issues and the resolution of multiple conflicts in inshore waters through the adoption of an integrated management approach (see Chapter 15).

In the meantime, scientific research has so far failed to resolve the public debate concerning the longer term consequences of intensive aquaculture. New concerns over the accidental escape of genetically modified farmed fish - notably salmon - and the implications for human health of high consumption levels of farmed fish simply add more fuel to the debate, as does the growing realisation of the high rates of exploitation of wildfish stocks needed to sustain the feed requirements of an expanding finfish aquaculture.

7. REFERENCES

Bailly, D. and Paquotte, Ph. (1996) Aquaculture and environment interaction in the perspective of renewable resource management theory, *Coastal Management,* **24**, 251-269.

Black, E., Gowen, R., Rosenthal, H., Roth, E., Stechy, D. and Taylor, F.J.R. (1997) The costs of eutrophication from salmon farming: implications for policy - a comment, *Journal of Environmental Management*, **50**, 105-109.

Bromley, D.W. (1991) *Environment and Economy: Property Rights and Public Policy*, Blackwell, Oxford.

Commission of the European Communities (1995) *Aquaculture and the Environment in the European Community*, Directorate-General XIV, Office for Official Publication of the European Communities, Luxembourg.

Edwards, V.M. and Steins, N.A. (1998) Developing an analytical framework for multiple-use commons, *Journal of Theoretical Politics,* **10**, 347-383.

European Commission (1999) *Statistical Bulletin, September 1999*, DG XIV, Brussels.

European Inland Fisheries Advisory Commission (1998) *Symposium on Water for Sustainable Inland Fisheries and Aquaculture, 20th session*, Praia do Carvoeiro, Portugal, 23 June - 1 July 1998.

FAO (2000) *The State of the World Fisheries and Aquaculture*, FAO, Rome.

Fingleton, P. and McCann, S. (1996) The use and regulation of coastal resources: traditional structures and modern national and EU overlays in western Ireland, *MARE Working Paper* No. 9, Aarhus University, Aarhus.

Goldburg, R. and Triplett, T. (1997) *Murky Waters: Environmental Effects of Aquaculture in the United States*, Environmental Defence Fund, Washington DC.

Holden, M. (1994) *The Common Fisheries Policy*, Fishing News Books, Oxford.

ICES Mariculture Committee (1998) *Report of the Working Group on Environmental Interactions of Mariculture*, International Council for the Exploration of the Sea, Copenhagen.

LEI-DLO, DIFER and SFIA (1997) *Property Rights in Fishing: Effects on the Industry and Effectiveness for Fishery Management Policy*, EU FAIR Programme on Property

Rights in the Fishing Industry, Landbouw-Economisch Instituut (LEI-DLO), The Hague.

Ross, A. and Roed, H. (2000) Aquaculture in Europe and its impacts on biodiversity, in L. Heaps (ed.), *Integrating Biodiversity and EU Fisheries Policy: Rebuilding Healthy and Productive Ecosystems, Final Report*, WWF - UK, Godalming, pp. 92-103.

Steins, N.A. (1998) 'We have to keep the foreigners out of our bay' - Top down regulations and the strategic response of Irish fishermen, in D. Symes (ed.), *Property Rights and Regulatory Systems for Fisheries*, Blackwell Science, Oxford, pp. 216- 229.

Stirling Aquaculture Ltd. (1996) The present state of aquaculture in the EU Member States and its future up to 2005: Executive Summary, in *Aquaculture: Development, Environmental Impact, Product quality Improvements*, Directorate-General for Research, STOA Programme, Brussels.

Symes, D. (ed.) (2000) *Fisheries Dependent Regions*, Blackwell Science, Oxford.

Van der Schans, J.W. (1996) Colonizing the coastal frontier: governing marine salmon farming in Scotland, in C. Baily, S. Jentoft and P. Sinclair (eds.), *Aquacultural Development: Social Dimensions of an Emerging Industry*, Westview Press, Boulder and Oxford, pp. 143-168.

Van der Schans, J.W. (1999) Governing aquaculture: dynamics and diversity in introducing salmon farming in Scotland, in J. Kooiman, M. van Vliet and S. Jentoft (eds.), *Creative Governance: Opportunities for Fisheries in Europe*, Ashgate, Aldershot, pp. 95-118.

Van Ginkel, R. (1991) *Elk Vist Op Zijn Tij: Een Zeeuwse Maritieme Gemeenschap, Yerseke 1870-1914*, Walburg Pers, Zutphen.

Weber, M.L. (1997) *Farming Salmon: A Briefing Book*, The Consultative Group of Biological Diversity, San Francisco.

Chapter 14

INSHORE FISHERIES, MARINE WILDLIFE CONSERVATION AND AN ECOSYSTEM BASED APPROACH TO MANAGEMENT

David Symes

1. INTRODUCTION

The inshore waters fringing the coastlines of Europe contain some of the most highly diverse marine ecosystems, in which man has evolved as the dominant species - a dynamic and disruptive predator. Concern has been growing for the quality of the marine environment and, in line with the ethos of the Biodiversity Convention (1992), the political rhetoric is switching its emphasis away from problems of fish stock sustainability *per se* to the wider and much more complex issues of ecosystem management and endorsement of the precautionary principle. As the longest established, most widespread and possibly most pervasive of human activities in these inshore waters, fishing has come under increasingly close scrutiny. Although initial attention was focused on the highly damaging effects of pollution in areas like the North Sea, the intensification of fishing effort, the development of new and more efficient technologies which seek to maintain yields even in the face of declining fish stocks, and the growth of industrialised forms of fishing have prompted the realisation that a good part of the ecological change is due to the persistence of fishing activity. Indeed, there is a general presumption that most, if not all, forms of fishing activity damage the marine ecosystem.

Concern over the loss of diversity, productivity and integrity of marine ecosystems has led to proposals for the closure of large areas of the sea to all forms of extractive activity in order to protect endangered habitats and ecosystems. Where such closures encompass offshore fishing grounds, the negative socio-economic effects, can, in some degree, be circumvented by the mobility of the offshore fishing fleets. But where they impinge upon

D. Symes and J. Phillipson (eds.), Inshore Fisheries Management, 239–256.

coastal waters, they are much more likely to impact seriously upon the less mobile, local inshore fishermen: for them, closure would mean not merely inconvenience but the loss of livelihood, social identity and cultural meaning.

It is important therefore for fishing and conservation interests to work in concert to achieve a balanced and well integrated system of management which not only conserves marine habitats and ecosystems but also protects inshore fishing interests. In effect, inshore fishermen must be seen as an integral part of the ecosystem and not sidelined or dismissed as an externality. The means by which this balancing of interests can be achieved is the concept of integrated management (Svelle *et al.*, 1997) wherein the objectives of sustainable fisheries and healthy marine ecosystems are considered jointly rather than separately. Decisions concerning fisheries must take full account of non-commercial elements in the ecosystem and decisions relating to the marine ecosystem must not set aside the interests of inshore fishermen.

The purpose of this chapter is to explore the relationships between fishing activities and marine conservation in the context of inshore fisheries in North West Europe. It starts with a brief description of inshore habitats and the impacts of fishing on their ecosystems, leading to a critical examination of alternative approaches to integrated management before finally examining some of the implications of a more radical ecosystem based approach to fisheries management.

2. INSHORE HABITATS

Inshore waters are highly productive systems: their nutrient supplies are enriched - and not infrequently over-enriched - from terrestrial sources through inflows of fresh water. The strong circulation of the inshore waters through coastal currents and tidal processes and - except where turbidity intervenes - the largely uninhibited processes of photosynthesis in shallow water ensure the basic productivity of the system. Where protected from vigorous wave action, well stratified sediments occupied by clearly defined benthic communities will develop. The more exposed inshore waters will, however, be vulnerable to periodic natural episodes - most commonly winter storms - which can severely damage the less stable habitats and the more sensitive ecosystems. And, of course, they are vulnerable to human intervention in the form of fishing and coastal aquaculture, sand and gravel extraction, oil and gas development, together with pollution from agricultural, industrial and urban waste water and from oil spills.

Table 14.1 Listed habitats and species (Natura 2000)

North Sea marine habitats, (Annex I of the Habitats Directive) whose conservation requires the designation of Special Areas of Conservation		**Seabirds and estuarine/coastal birds occurring around the North Sea, listed in Annex I of the Birds Directive**	
Estuaries		Red throated diver	*Gavia stellata*
Lagoons		Black throated diver	*Gavia arctica*
Large shallow inlets and bays		Great northern diver	*Gavia immer*
Submerged or partly submerged sea caves		Slavonian grebe	*Podiceps auritus*
Sandbanks slightly covered by seawater at all time		Storm petrel	*Hydrobates pelagicus*
		Leach's petrel	*Oceanodromo leucorhoa*
Mudflats and sandflats not covered by seawater at low tide		Avocet	*Recurvirostra avosetta*
		Golden plover	*Pluvialis apricaria*
Reefs		Red necked phalarope	*Phalaropus lobatus*
		Mediterranean gull	*Larus melanocephalus*
		Sandwich tern	*Sterna sandvicensis*
North Sea marine species, (Annex II of the Habitats Directive) whose conservation requires the designation of Special Areas of Conservation		Roseate tern	*Sterna dougallii*
		Common tern	*Sterna hirudo*
		Arctic tern	*Sterna paradisaea*
		Little tern	*Sterna albifrons*
		Black tern	*Chlidonias niger*
Grey seal	*Halichoerus gruppus*		
Common seal	*Phoca vitulina*		
Bottlenose dolphin	*Tursiops truncatus*		
Harbour porpoise	*Phocoena phocoena*		
Otter	*Lutra lutra*		
Allis shad	*Alosa alosa*		
Twaite shad	*Alosa fallax*	[Note: there is a more extensive list of resident and migratory seabirds and estuarine/coastal birds not included in Annex I for which the North Sea and its margins constitute important breeding or overwintering areas].	
Lampern	*Lampetra fluviatitis*		
Sea lamprey	*Petromyzon marinus*		
Sturopen	*Acipenser sturio*		
Houting	*Coregonus oxyrinthus*		

These inshore waters comprise a variety of habitats including both inter-tidal and sub-tidal areas. Annex I of the EU's Habitats Directive (92/43/EEC) has identified seven marine habitats in need of conservation, together with a list of threatened species (Annex II), covering benthic fauna, commercial and non-commercial fish and sea mammals (see Table 14.1). A further list of endangered seabirds is included in Annex I of the earlier Directive on the Conservation of Wild Birds (79/109/EEC). The listed habitats are not discrete entities: they occur in broadly overlapping sets which can for convenience be arranged in structured form (Table 14.2). Apart from sea caves and reefs, they are features characteristic of low lying coasts with extensive inter-tidal areas forming a dominant fringing zone.

Three broad physiographic types can be identified:

- *estuaries*, characterised by brackish tidal waters exhibiting strong freshwater influences, carrying relatively heavy sediment loads deposited in slack water on the margins as sand- or mud-flats and, at

Table 14.2 Inshore waters: key habitats, target species and fishing gears

Coastal habitats		Target species	Fishing gears	
(i) Physiographic form	(ii) Fringing habitat		(i) Fixed gears	(ii) Mobile gears
Estuaries (brackish water)	mud/sandflats	shellfish (molluscs), bait	-	handraking mechanical/suction dredges
	lagoons	[little fishing activity in temperate waters]		
	open waters	eels, salmon, seatrout	gillnets, fyke nets	-
Large shallow inlets and bays (salt water)	mud/sandflats and submerged sandbanks	shellfish (molluscs)	-	handraking mechanical/suction dredges
	lagoons	[little fishing activity in temperate waters]		
	reefs	crustaceans (crabs, lobsters) demersal fish	pots	spring loaded dredges rock hopper trawls
	open waters	crustaceans (shrimps) demersal fish	gillnets trammel nets	trawls
Cliffs	reefs	crustaceans (crabs, lobsters) demersal fish	pots	spring loaded dredges rock hopper trawls
	open waters	demersal fish pelagic fish crustaceans (*nephrops*)	hand/longlines gillnets	otter/beam trawls purse seines trawls

their mouths, as deltas; the flats frequently provide important feeding grounds for overwintering and passage migrant waders and wildfowl;

- *shallow inlets and bays* with water depths rarely exceeding 30 m, also characterised by depositional features (fringing sand- or mud-flats and sub-littoral sandbanks) with a diversity of sediments and substrates and a well developed zonation of benthic communities and high biodiversity (Gubbay, 1997);

- *cliffed coastlines* usually formed of exposures of hard rock (but occasionally much softer materials e.g. boulder clays); submerged or partly submerged sea caves are identified in the list of habitats to be protected under the 1992 Directive, as are the submarine reefs formed of rock substrates or biogenic concretions and commonly - though not exclusively - associated with cliffed coastlines. The seacliffs themselves may be important as breeding sites for a range of seabirds.

3. IMPACTS OF INSHORE FISHING ACTIVITY

In assessing the effects of fishing activity on marine habitats and ecosystems, we need to bear in mind that their existing status is, in very many cases, the product of a very long history of exploitation through fishing, often stretching back to prehistoric times. There are probably no areas of inshore water around the coasts of Europe which are unaffected by past or present fishing activity. On certain fishing grounds intensive levels of exploitation have been continuing for well over a century. Most of the significant changes to the natural ecosystems occurred in the early stages of exploitation and the impacts of continuing exploitation through fishing may be quite marginal (Jennings and Kaiser, 1998). In almost all instances it is simply not possible to restore such heavily exploited ecosystems to anything approaching their original pristine state. It may, however, be feasible to increase their biodiversity and enhance their natural productivity through carefully controlled levels of future fishing activity. Moreover, the damage caused by continuing fishing activity is probably far less than that which can be expected from the effects of winter storms or long term marine climatic change and sea level rise.

Notwithstanding these caveats, there is abundant evidence to suggest that fishing will continue to exert a damaging, if not critical, impact on marine habitats and their ecosystems (Jennings and Kaiser, 1998). The increasing size and weight of modern fishing gears, such as beam trawls, and the attachment of devices to expose the target species buried in surface sediments, will inevitably disturb the habitat and crush the more fragile forms of benthic epifauna and infauna. Recovery from physical damage and loss of benthic fauna may be quite rapid - weeks rather than years - but where recovery is slow the benthic communities may be permanently altered through the colonisation of short lived but opportunistic scavenging species. Of particular significance for inshore fisheries are the modern methods used in fishing for cockles, mussels, oysters and scallops. Mechanical dredges - and more particularly the suction dredges used in cockle fishing - will churn up the sediments, threaten erosion of the sandbanks and remove up to 80% of the benthic fauna (Gubbay, 1997).

Popular concern over the effects of fishing has focused principally on the incidental capture of marine mammals. Five species - grey and common seal, bottle nose dolphin, harbour porpoise and otter, all included in the Habitats Directive's list of species requiring conservation - risk entrapment in fixed nets, or, in the case of the otter, in crab and lobster pots as well as fyke nets. Although the images are distressing, only in one case - the harbour porpoise - is the bycatch a matter of concern for the maintenance of population numbers. An estimated 7,000 harbour porpoises are killed annually in gillnets set by Danish fishermen in the North Sea, mainly in offshore waters (Vinther, 1994). Set against the recent decline in its regional population, the evident ageing of the harbour porpoise population and its inherently low fecundity, these figures suggest that the harbour porpoise population may be unable to withstand such a level of sustained incidental mortality (Woodley and Read, 1991). Seabirds, especially the deep diving gannet and auks, are also at risk from entanglement in fixed nets set close to the shore during the nesting season, while some surface scavenging gulls, together with the gannet, are at risk from longlining mainly in offshore waters (Dunn, 1998).

Some of the most significant changes brought about by intensive fishing activity occur in the structure of the fish populations themselves. The most easily observed effects concern the mortality and spawning stock biomass of the target species. In the North Sea, for example, it is estimated that between 30 and 40% of the total biomass of commercially exploited species is removed each year (Gislason, 1994). For cod and haddock, the rates may be as high as 50%. Yet the total productivity of the North Sea ecosystem seems unimpaired. There are few reported 'crashes' among the commercial fish populations in the region - mackerel in the 1960s and herring a decade later may be among the exceptions - and biological extinctions among non-commercial species are also rare. Where they have occurred it is impossible to lay the blame solely on the effects of fishing.

Changes will, however, occur in species composition and size structure of the overall population. Heavy fishing pressure on the top predator species, principally cod and haddock in the North Sea, will almost inevitably create opportunities for relative increases in the smaller prey species which may set in train further alterations to the food systems within the ecosystem.

As man extends his role as a top predator within the marine ecosystem, so new threats to the overall balance of the system are created. The latest problem is one which is generated by the offshore, rather than the inshore, fishing sector but which has serious implications for the coastal ecosystem. It concerns the massive expansion of the industrial fisheries for Norway pout, sprat and, in particular, sandeel. In some ways, the very development of the fishery testifies to the man-induced changes to the structure of the North Sea ecosystem with the huge biomass of these small fish reflecting the reduction in numbers of their principal predators through fishing. But these small fish are also vital elements in the complex food web which supports the ecosystem of the North Sea and in particular the diet of a wide range of seabirds, large aggregations of which occur in breeding colonies around the North Sea coast. The breeding sites are chosen in relation to reliable and abundant sources of food, including the large populations of sandeels which make up around one third of their total food consumption and just less than one half during the breeding season (Tasker and Furness, 1996). The availability of this primary food source would be threatened by any further expansion of the industrial fishery and its incursion into coastal waters close to major breeding colonies, as in the case of the Wee Bankie Fishery close to the Isle of May on the east coast of Scotland.

The estuaries and low lying coasts, fringed by extensive sandbanks and mudflats, around the Atlantic and North Sea also provide over-wintering sites or vital feeding grounds for large numbers of migrating waders and waterfowl. But their ability to sustain this function is under threat from overexploitation of shellfish stocks (mussels, cockles, oysters and clams) which also provide a major source of food for the bird populations, and by damage or disturbance to the habitats through the use of mechanical or suction dredges. Even the basic activity of bait digging can adversely affect the feeding behavior of the birds (Fowler, 1992).

Not all fishing activity has a negative impact upon marine wildlife populations. Indeed, the much criticised practice of discarding fish at sea - together with the dumping of offal produced by primary processing on board - has, in fact, made a significant contribution to supplies of food for the seabird populations. Offals and discards account for *circa* 30% of the total food consumed by seabirds in the North Sea and around half the food consumed in winter (Tasker and Furness, 1996). Several surface scavenging gulls have recorded substantial increases in their populations over the past 50 years or so; those increases are at least in part attributable to increases in waste by-products of the modern fishing industry.

4. INTEGRATED FISHERIES MANAGEMENT AND MARINE WILDLIFE CONSERVATION

4.1 Integrated fisheries management and the precautionary principle

There can be little doubt that existing policies have failed to achieve a sound basis for the management of either sustainable fisheries or a healthy marine ecosystem. In inshore fisheries, ill defined property rights, inadequate systems of regulation and the 'disembedding' of management from the local community are variously blamed for the inability to restrain increases in fishing effort and to prevent the incursion of inappropriate fishing practices within inshore waters. Whereas locally constructed systems for the management of inshore fisheries were necessarily predicated on the premise of long term sustainability and equity - simply because the inshore fishermen had nowhere else to go - it has usually proved possible for the more mobile offshore sector to escape the consequences of localised failures in a state led regulatory system by moving elsewhere. 'Integrated fisheries management' and the implementation of the precautionary principle are seen by some as providing a new opportunity to establish the basis for the sustainable use of the marine ecosystem and perhaps a last chance to ensure the long term viability of the inshore fishing industry.

According to the UN Convention on Environment and Development 1992:

> 'In order to protect the environment, the precautionary approach shall be widely applied by states according to their capabilities. Where there are threats of serious or irreversible damage to the environment lack of full scientific certainty should not be used as a reason for postponing cost effective measures to prevent degradation'.

This somewhat imprecise assertion of the precautionary principle has led to several distortions of meaning when applied to fishing. Within fisheries management, its application is seen to provide a more robust mechanism for the rational exploitation of fish stocks, setting both target and threshold values to ensure that stocks remain within safe biological limits (Gislason and Kirkegaard, 1998). However, this does not necessarily mean that fishing remains within reasonable bounds as far as the diversity, productivity and integrity of the wider ecosystem is concerned. Conversely, interpretations of the precautionary principle by some conservationists are used to buttress arguments for the closure of certain fishing grounds on the pretext that damage - often unspecified and unquantified - is being done to the marine environment and unproven assumptions that a moratorium on fishing will help to restore the ecosystem to something akin to its earlier undamaged state. Correctly interpreted and sensibly used, however, the precautionary principle can operate as an essential bridging device linking both fisheries management and marine wildlife conservation within a broader, overarching concept of integrated management.

These two concepts - the precautionary principle and integrated management - have made considerable political gains in recent years. The European Community formally embraced the precautionary principle in 1992:

> 'Community policy on the environment... shall be based on the precautionary principle... (and) environmental protection requirements must be integrated into the definition and implementation of other Community policies' (Maastricht Treaty, Article 130 (2))

More specifically within fisheries, the principle was elaborated within the FAO's *Code of Conduct for Responsible Fishing* (1995); and the concept of integrated management was endorsed by the Intermediate Ministerial Meeting on the Integration of Fisheries and Environmental Issues held in Bergen in 1997 as the proper basis for the conservation of the North Sea ecosystem.

But so far the intention has not been translated into appropriate regulation. Apart from the recent decision to ban driftnet fishing for tuna in European Union waters, there has been little to indicate a shift of emphasis in the Common Fisheries Policy towards a precautionary or more integrated approach.

Action has also recently been taken to contain the threat to seabird populations on the north and east coasts of Scotland from the relatively uncontrolled sandeel fisheries. In 1998 a local agreement was reached between the Royal Society for the Protection of Birds and the Shetland Fish Producers' Organisation in respect of the sandeel fishery off Shetland concerning the level of catch, a precautionary closure during the chick-rearing summer months (June-July) and a limitation on the size of boats. The significance of this example is that a precautionary local management strategy was put in place as the result of direct negotiations between the principal stakeholders; an earlier attempt in 1995 by the Scottish fisheries department to balance the interests of the environmental lobby and the fishing industry had failed to satisfy either party (Dunn and Goodlad, 2000). Off the east coast of Scotland, the European Commission in 1999 introduced a seasonal ban on sandeel fishing in the area of the Wee Bankie, close to the Isle of May kittiwake colony, for a trial period. In this case it had been suggested that the breeding success of the kittiwake colonies could be used as a surrogate measure of the viability of the marine food web in the northern North Sea; thresholds of breeding success could be used to trigger the closure and reopening of the sandeel fisheries.

As signatory to the Biodiversity Convention, the EU is required to formulate a strategy to 'anticipate, prevent and attack the causes of significant reduction or loss of biological diversity' (Article 6). The EU's Biodiversity Strategy was published in 1998, in which it identified the need to develop sectoral (and cross-sectoral) Action Plans for all relevant policy areas, including fisheries by 2000. Early drafts of the plan for fisheries suggest that the Commission has some way to go before providing a convincing framework for the realisation of the basic aims of the Biodiversity Convention in the context of Europe's fisheries.

There is perhaps rather more evidence of an integrated approach to fisheries management at Member State level in respect of their territorial waters within the 6 and 12 mile limits. Examples include the international protocol involving Denmark, Germany and the Netherlands for the joint management of the Wadden Sea and the changes to national legislation in the UK in order to implement the Habitats Directive (Symes and Phillipson, 1997) and to expand the role of local Sea Fisheries Committees to include obligations in respect of environmental conservation. There may well be some divergence and a potential point of conflict between the willingness of the coastal state to engage in meaningful integrated management within the 6-12 mile zone and the need for a regional ecosystem approach, covering an area like the North Sea, which would require a somewhat reluctant EU to take the lead. Ideally, of course, the regional and zonal approaches need to be very closely coordinated to maximise the benefits of integrated management for both fisheries and the marine ecosystem.

4.2 Designated areas: no take zones

For some the concept of managing an ecosystem is a contradiction in terms: the notion of human control over a natural ecosystem is incongruous. In practical terms, the problems of managing a marine ecosystem are made more difficult by the fluidity and permeability of its boundaries, the absence of clearly defined political responsibility and a lack of hard scientific knowledge and understanding of marine ecosystems and the impacts of human intervention through fishing. There are difficulties in defining the objectives of ecosystem management: while the general aims of maintaining the diversity, productivity and integrity of an ecosystem present no problems, the challenge for ecologists lies in specifying targets for the ecosystem, to set alongside those established for the commercial fisheries, and in operationalising such objectives. Where 'full scientific certainty' is missing, it may seem all too easy to circumvent the definition and operationalisation of objectives by choosing the easy option of closing the fishery and applying the principle of 'passive management' by removing all potentially disruptive sources of human intervention.

Designated areas are much more difficult to implement in marine areas, where there are no easily observed natural boundaries. The tendency therefore is to use geometric boundaries (e.g. ICES subdivisions) which can be defined on sea charts and monitored through satellite technology. Designated areas in the form of marine protection areas (MPAs), marine national parks (MNPs) and, marine nature reserves (MNRs) are perceived as the optimal form of marine ecosystem management by some conservation interests enabling the regulation of fishing activity through zoning and where necessary, the banning of fishing altogether.

Because of the edge effects of a designated area superimposed on an unbounded ecosystem and the risks of contamination from activities outside the closed area and the leakage of benefits into neighbouring areas, no take zones in temperate waters would have to cover very large areas of sea to achieve their goals - up to 25% of the North Sea, for example (Garrod, 1997) - and remain in place for 20 years or more. Such a system would almost certainly envelop some of Europe's most important fishing grounds and, depending on the precise location of the closed areas, encroach upon the inshore fisheries either directly or indirectly. The assumed benefits for the sustainability of the ecosystem would no doubt be considerable (Roberts, 1998). But so too would be the economic and social costs to the fishing industry and to the local community in fishing dependent regions.

It is at least debatable as to whether similar results could be achieved from less draconian action. Permanent closure of inshore waters would be disastrous for the neighbouring fishing communities. An alternative approach could be sought in the implementation of much stricter regulations in respect of

- the number and size of vessels permitted to fish in a given area;
- the number of fishing days allowed;
- the banning of certain fishing gears known to inflict critical damage on sensitive habitats (e.g. beam trawls and suction dredges);
- the specification of permitted gears;
- the use of seasonal closures to protect breeding and nursery grounds;
- the introduction of emergency closures where, for example, the proportion of juvenile fish exceeds a given level; and
- a system of 'ecological impact assessments' conducted prior to allowing any new form of fishing activity.

Most of these regulatory mechanisms are already applied in the management of inshore waters, as for example in the byelaws governing fishing in the Sea Fisheries Districts in England and Wales. Sadly even this rich panoply of measures is not always sufficient to guarantee the sustainability of the inshore fisheries or their ecosystems (Symes and Phillipson, 1998).

4.3 Natura 2000

Direct action in furtherance of the aims of marine conservation in European inshore waters has, prior to implementation of the Habitats Directive, met with only very modest success. In the UK, for example, proposals for designated areas under the provisions of the *Wildlife and Countryside Act 1981* - in the form of Marine Nature Reserves (MNRs) - resulted in only three designations and none in the North Sea region. Failure to establish more MNRs was due largely to protracted negotiations between conflicting interests - including the fishing industry - and an inability to reach meaningful compromise (Symes and Phillipson, 1997). A much weaker line of defence for the conservation of important marine sites is provided by the non-statutory designation of 29 Marine Consultation Areas in Scotland, where Scottish Natural Heritage has the right to be consulted over development proposals including fish farms, and 27 Sensitive Marine Areas in England and Wales where informal guidelines are to be developed for the control of human activity. Otherwise, marine nature conservation has come to depend much more upon voluntary agreements and the setting up of 15 non-statutory reserves, only three of which are located on the North Sea coast.

However, there is a recent development which could make a more significant contribution to the integrated management of inshore waters. *Natura 2000* is an EU initiative which aims to establish a coherent network of special conservation areas throughout the Community by means of the designation and management of Special Protection Areas (SPAs) under the Birds Directive (74/409/EEC) and Special Areas of Conservation (SACs) as part of the Habitats Directive (92/43/EEC). The two Directives are complementary. The earlier Birds Directive provides for the protection, management and control of all species of naturally occurring wild birds and, in respect of some 175 listed species, establishes the need for adequate measures to preserve a sufficient diversity of habitats and maintain populations at ecologically acceptable levels. The Habitats Directive will contribute to maintaining biodiversity by requiring Member States to adopt measures to maintain or restore threatened natural habitats and conserve species which are considered endangered, vulnerable or rare at a number of key sites throughout the Community.

The vast majority of designated sites under the two Directives are land based, but the significance of the Habitats Directive is that it provides for new designations in marine areas. Even though the Annex I list of birds requiring special conservation action includes a significant number of seabirds, waders and wildfowl, SPA sites - in the UK at least - are confined to terrestrial and inter-tidal areas, building on the system of existing designations of Sites of Special Scientific Interest (SSSIs). By contrast, SACs are to be established in both terrestrial and marine environments, including estuaries, tidal mudflats and sandflats, large shallow bays and inlets, lagoons and submerged sandbanks and reefs. Clearly, the initial intention of the Directives was that they would be confined to coastal waters, that is within the 6 or 12 mile limits.

Progress towards the selection of candidate SACs has been patchy. The final lists of proposed sites were due to be submitted to Brussels in June 1998 with a view to their approval by the end of that year and their full implementation, including the formulation of management plans, in 2004.

There have been criticisms in some countries that the initial selection of sites was inadequate and a number of new sites have been added since the original deadline. In UK waters, for example, the number of confirmed and proposed sites currently stands at over 40, compared with the initial list of 35. Also, following a landmark judgement in the UK High Court which ruled that a coastal state's responsibility for protecting the marine environment extended beyond its 12 mile limits it is now likely that further SACs will be designated in offshore waters.

Designation of marine SACs involves the creation of a new management structure involving local authorities and interest groups, the definition of specific management objectives and the development of a detailed management plan. According to Article 6(2) of the Directive

> 'Member States shall take appropriate steps to avoid, in the special areas of conservation, the deterioration of natural habitats and the habitats of species as well as disturbance of the species for which the areas have been designated ...'

In the implementation of the management plan, securing this objective will largely depend on the principle of voluntary compliance with its aims and objectives, and only in exceptional circumstances will new legislation be used to regulate existing activity, including fishing. It is therefore unlikely that SACs will have severe negative repercussions for existing inshore fisheries. However, any new development project - including presumably, the development of new fisheries or the use of new fishing methods - will be subject to an appropriate scientific assessment to ensure that it will not adversely affect the integrity of the site or the viability of the listed species. In this way, the precautionary principle will be correctly applied.

5. INSHORE FISHERIES AND AN ECOSYSTEM BASED APPROACH TO MANAGEMENT

There is rather more to the ecosystem approach than simply a closer coordination of fisheries management and marine wildlife conservation. With its origins in the Biodiversity Convention in 1992 and the concern for the sustainability of the biodiversity, productivity and integrity of terrestrial and marine ecosystems, the ecosystem approach represents a step change from the earlier attempts to bridge the gap between a fisheries centred, sectoral management policy, on the one hand, and attempts to protect threatened habitats and species from the impacts of fishing through the designation of sites of conservation interest, on the other. The concept is still in its infancy and has not yet been resolved into a simple, unambiguous definition, though its principles have been identified - the so-called Malawi principles - and adopted.

The absence of a clear working definition has led some of its critics to invest the concept with exaggerated ambitions, bringing it dangerously close to the notion of 'ecosystem management' linked to socially constructed views of 'ideal ecosystems' usually construed as in a pristine state. Man is not yet - and probably never will be - in a position to provide a blueprint for the ideal ecosystem; he lacks sufficient knowledge and understanding of the essential functions, processes and interactions that govern marine ecosystems. Nor has he the power to intervene constructively to modify those functions, processes and interactions in pursuit of some socially constructed model of the ideal ecosystem. At the same time, the insistence that the ecosystem approach incorporates man as a keystone species has led others falsely to portray the ecosystem approach as embracing economic, social and cultural objectives in addition to those concerned with the sustainability of biodiversity. As a consequence of these dangerous over-elaborations, the ecosystem approach is sometimes depicted as incapable of offering practical guidance to managers (Larkin, 1996).

In practice, the ecosystem approach is driven by more modest ambitions and its emphasis is utilitarian rather than ecocentrist, seeking to balance the intrinsic and applied values of the ecosystem. It aims to manage the habitats and populations that make up the marine ecosystem in mutually beneficial ways that will secure the sustainable development of commercial and non-commercial species alike. Thus the focus of management is shifted away from the single species approach to both fisheries and wildlife conservation and towards ensuring, as far as possible, that the ecosystem as a whole does not sustain irretrievable damage. The by-product of this approach should be long term sustainable yields from commercial stocks. As we cannot know what values will be attached to marine ecosystems by future generations, it is imperative that today's managers take as broad a view of the ecosystem as possible, coupled with a strongly precautionary approach.

Because of our relatively undeveloped knowledge of the functional relationships and processes occurring within marine ecosystems, at present only a pragmatic and provisional ecosystem approach is possible. According to Pope and Symes (2000) this could be based on four operational objectives:

- informing fisheries managers of the ecosystem consequences of particular management options in order to ensure that actions which may damage the ecosystem are not undertaken as a result of ignorance;
- allowing fishing to affect the wider ecosystem adversely only where it is consistent with the rational and responsible exploitation of target stocks;
- ensuring the fullest protection for essential habitats for commercial and non-commercial species; and
- designating ecosystem limit reference points for commercial and non-commercial species to ensure that no component species falls below safe biological levels.

Further research is needed to bring us to the point where we can begin to develop a more sophisticated ecosystem approach based on detailed understanding of the marine ecosystem's functioning.

Although the adoption of an ecosystem approach could signal a period of significant restructuring for the fishing industry in general, it would not necessarily mean any additional hardship for the sector beyond that necessary to guarantee the recovery of commercial fish stocks. Indeed, it is precisely because draconian measures are now needed to restore the balance between harvesting capacity and depleted resources throughout much of the north east Atlantic, that an opportunity arises to broaden the basis of fisheries management by including the elements of an ecosystem approach outlined above. Among the measures required are first and foremost a significant reduction of fishing capacity through decommissioning up to 40% of the existing fishing fleet; strengthening technical conservation measures to ensure the protection of essential habitats and spawning, nursery and juvenile populations of commercial and non-commercial species; and restructuring grant aid to encourage the adoption of ecologically responsible fishing practices (Symes and Pope, 2000). Specifically in relation to inshore fisheries, this could involve ending the neutral stance towards capacity reductions in the under 12 m sector in Multi-Annual Guidance Programmes; introducing, through licensing, exclusive access rights to inshore waters for locally registered small boats; banning certain fishing methods in areas of high ecological sensitivity; and reassessing area and seasonal closures to ensure coherence with the principles of essential habitat protection. There will be a particular need to ensure that measures to curb overcapitalisation in the offshore sector do not lead to increased pressures on inshore waters through the redeployment of displaced fishing effort or through the reinvestment of proceeds from decommissioning in the inshore sector.

A key to the ecosystem approach is the development of ecosystem management plans, partly to serve as baseline assessments of the marine ecosystem but more particularly to identify limits for the ecosystem impacts of fishing activity and articulate medium or long term strategies for fisheries development. Ecosystem management plans will also provide a more structured framework for existing and future actions concerned with marine wildlife conservation. To date, there are very few examples of comprehensive management plans for fisheries in Europe, though since the passing of the Magnussen-Stevens Act in 1976 they have been a feature of fisheries management in the USA. Here proposals have recently been laid before Congress to require the regional authorities to develop ecosystem management plans to set alongside existing fisheries management plans (National Marine Fisheries Service, 1999). Parallel developments in Europe may well have to await decisions to decentralise the Common Fisheries Policy and create regional management organisations at the level of the regional seas (see Symes, 1998).

Drafting regional ecosystem management plans would require the reintegration of inshore fisheries within the wider regional approach: inshore management plans would thus become a subset of the regional plan. In practice, however, political reality suggests that we may have to settle for a more fragmented approach within inshore waters where the individual coastal state remains the responsible authority. Providing all coastal states are working to the same broad objectives, this situation may not be without advantage for it will allow detailed management strategies to be varied in accordance with both local ecological conditions and fishing practice.

6. REFERENCES

Dunn, E. (1998) The impact of fisheries on sea birds in the North Sea, in D. Symes (ed.), *Northern Waters: Management Issues and Practice*, Blackwell Science, Oxford, pp. 208-215.

Dunn, E. and Goodland, J. (2000) The Shetland sandeel fishery: consensus rather than conflict, in L. Heaps (ed.), *Ecosystem Approach and Integrating Environmental Considerations into EU Fisheries Policy*, Report of Workshop held at Vilanova i la Geltris, 18-19 May, 2000, WWP, London, pp. 34-36.

Food and Agriculture Organisation (1995) *Code of Conduct for Responsible Fisheries*, FAO, Rome.

Fowler, S.L. (1992) *Survey of Bait Collection in Britain*, JNCC Report no. 107, Joint Nature Conservation Committee, Peterborough.

Garrod, D.J. (1997) Where do we go now to conserve fish stocks? Paper presented at *Greenwich Forum Conference on Rethinking the Common Fisheries Policy - Looking Towards 2002*, London, 16 October 1997.

Gislason, H. (1994) Ecosystem effects on fishing activities in the North Sea, *Marine Pollution Bulletin*, **29**, 520-527.

Gislason, H. and E. Kirkegaard (1998) Is the industrial fishery in the North Sea sustainable? in D. Symes (ed.), *Northern Waters: Management Issues and Practice*, Blackwell Science, Oxford, pp. 195-207.

Gubbay, S. (1997) *Generic Effects of Fishing on Natura 2000 Habitats and Species*, report to the Joint Nature Conservation Committee, JNCC, Peterborough.

Jennings, S. and M. Kaiser (1998) The effects of fishing on marine ecosystems, *Advances in Marine Biology*, **34**, 201-351.

Larkin, P.A. (1996) Concepts and issues in marine ecosystem management, *Reviews in Fish Biology and Fisheries*, **6**, 139-164.

National Marine Fisheries Service (1999) *Ecosystem-Based Fisheries Management*, A Report to Congress by the Ecosystem Principles Advisory Board, Department of Commerce, Washington.

Pope, J. and Symes, D. (2000) *An Ecosystem Based Approach to the Common Fisheries Policy: Defining the Goals*, English Nature, Peterborough.

Roberts, C. (1998) No-take marine reserves: briefing notes, paper tabled at *WWF Workshop on No-Take Zones: The Way Forward*, London, 7-8 April 1998.

Svelle, M., Aarefjord, H., Heir, H.T. and Øverland, S. (1997) *Assessment Report on Fisheries and Fisheries related Species and Habitats Issues*, Ministry of the Environment, Oslo.

Symes, D. (1998) Towards 2002: subsidiarity and the regionalization of the Common Fisheries Policy, in T. Gray (ed.), *The Politics of Fishing*, Macmillan, London, pp. 176-193.

Symes, D. and Phillipson, J. (1997) Inshore fisheries management in the UK: Sea Fisheries Committees and the challenge of marine environmental management, *Marine Policy* **21** (3), 207-224.

Symes, D. and Phillipson, J. (1998) Fishing within limits: inshore fisheries management and the concept of local preference, in D. Symes (ed.), *Property Rights and Regulatory Systems in Fisheries*, Blackwell Science, Oxford, pp. 201-215.

Symes, D. and Pope, J. (2000) *An Ecosystem Based Approach to the Common Fisheries Policy: Achieving the Objectives*, English Nature, Peterborough.

Tasker, M. and Furness, R.W. (1996) Estimation of food consumption by seabirds in the North Sea, in G.L. Hunt and R.W. Furness (eds.), *Seabird/Fish Interactions, with Particular Reference to the North Sea*, ICES Cooperative Research Report No. 216, ICES, Copenhagen, pp. 6-42.

Vinther, M. (1994) Incidental catch of harbour porpoise *(Phocoena phocoena)* in the Danish North Sea gill net fisheries: preliminary results, *Scientific Symposium on the 1993 North Sea Quality Status Report*, Ebbettoft, Denmark, 18-21 April.

Woodley, T.H. and Read, A.J. (1991) Potential rates of increase of harbour porpoise (*Phocoena phocoena*) population subjected to incidental mortality in commercial fisheries, *Canadian Journal of Fish Aquatic Science*, **48**, 2429-2435.

Chapter 15

MULTI-USE CONFLICTS IN INSHORE WATERS

Rob van Ginkel and Nathalie Steins

1. INTRODUCTION

Until only quite recently, human activity in Europe's inshore waters made little impact on marine resources and one activity scarcely interfered with any other. But long term developments such as demographic growth, urbanisation, expanding demand for food and natural resources, the integration of resources into markets, technological innovation, the development of marine transport and the rise of tourism have all increased the level of human impact on coastal areas and resources. Once a domain where only fishers and sailors ventured, the sea is now being used for many other purposes, particularly in the inshore zone. In addition to their continuing importance for fisheries, inshore waters are now commonly used quite intensively for other potentially intrusive activities. The public's growing demand for entry to the inshore zone has brought about competition and conflicts between various claimants over access to, allocation of and control over coastal space and resources. In addition to multiple use conflicts, intensified use of the coastal zone can also exacerbate resource management problems. Human activities unrelated to fisheries affect marine ecosystems: mineral exploration and development, water based recreation, navigation, dredging, land reclamation, industrial and agricultural waste disposal including the discharge of effluents and the dumping of toxic chemical and nuclear waste, thermal pollution from coastal power stations etc.

The present chapter outlines some of the issues concerning the diversified and intensified use of inshore waters. The coastal zone is a typical example of a common pool resource, where exclusion of access is difficult and joint use involves subtractability. It is clear that as the number of resource users

D. Symes and J. Phillipson (eds.), Inshore Fisheries Management, 257–273.

and the types and extent of resource use multiply, interdependencies increase. Obviously, there are conflicts of interest when activities interact. Therefore, 'when commons evolve into multiple use resources, the institutional framework within which collective resource use takes place has to be re-negotiated to avoid adverse impacts associated with increased access of new users to the resource system, such as overexploitation, alienation of traditional users and inter-user conflicts' (Steins and Edwards, 1999: p. 242). However, we are dealing with complex situations where different uses may be mutually incompatible; where one activity may be damaging to some interests but beneficial to others; where some activities cause effects elsewhere and are themselves affected by different activities; where others cause effects without being interfered with themselves; where still other activities only suffer interference without causing any effects on others; and where different user groups are regulated through different management regimes (Sibthorp, 1975; van Hoorn *et al.*, 1980; Edwards and Steins, 1998). In this chapter, we focus only on how inshore fishing is affected by the utilisation of the same space and by other activities occurring in the coastal zone. Attention is not paid to the other side of the coin - how fishing impacts on other activities. This lop-sided approach is to some extent justified by the fact that it is usually the fishing industry that is forced to retreat under pressure of development by other users of the coastal zone, whether for reasons of relative organisational strength, public opinion or governmental preference. The one area where fisheries are perceived to have potentially strong negative impacts is the marine environment and the concomitant concern for marine wildlife conservation, already considered in Chapter 14.

We usually speak of resource management when referring to situations in which people exploit common pool resources under certain management regimes. However, it is rarely the resources *per se* that are managed; management usually relates to geographical space and human behaviour. Spatial management refers to the locations where particular forms of resource use are allowed. People management relates to the question of which categories of individuals or groups are entitled to use which resources, where and when. Here, exclusion mechanisms restrict the number of people who are allowed access to the marine space and the resources in it. Thus, the management of multi-use situations relates to three complementary dimensions: resource management, spatial management and people management. All three dimensions are potentially conflictual at local, regional, national and international levels. But multi-use conflicts are first and foremost conflicts between different user groups over the consumption of space. Thus, solutions with respect to competing uses typically involve segregation of use over space and/or time.

2. INSHORE FISHERIES: THE IMPACTS OF OTHER FORMS OF MARINE SPACE UTILISATION

A wide variety of other activities affect fishing directly or indirectly. In this section we examine how different categories of use impact on inshore fisheries beginning with the case of how one type of fishery may interfere with another and, recapping briefly on results from Chapter 13, how marine aquaculture can be at odds with established fishing interests, before moving on to consider the repercussions of extractive and non-extractive uses and the case sometimes made for the non-use of marine resources and finally to assess the impacts of the abuse of the marine environment for purposes of waste disposal on inshore fisheries.

2.1 Capture fisheries versus capture fisheries

Although fishers may compete with each other for the same species or exploit the same ecosystem, they rarely exploit only one species or one ecological niche. Normally they catch different species with different kinds of gear over an annual cycle, using various ecological niches. Switching behaviour is 'the most important adaptive strategy used by fishermen' (Acheson, 1988: p. 49). The availability of alternative niches and resources is, of course, a precondition for this kind of flexibility. The extraction of renewable marine resources by similar categories of users may lead to congestion, while the uses of different fishing gears - for example, stationary versus mobile gear - in any one particular location may be incompatible. Disputes over space are usually zero-sum games in which there are both winners and losers. As Jentoft and Kristoffersen (1989) point out in relation to the spatial management of the Lofoten cod fishery, making more space available on the fishing grounds for one type of gear simply reduces the space available for other gears. In the French Mediterranean conflicts arise between different components of the inshore fleet relating to competition for resources and space. This applies in particular to situations where mobile gear and stationary gear are deployed in the same waters and also when there are illegal incursions into the inshore zone where fixed gears are used (Farrugio and Le Corre, 1991; Dufour, 1996). Similar problems occur in Greece, though here time and area closures are imposed to limit gear interaction and fishing effort. According to Frangoudes (1996: p. 124) the decisions are usually 'the result of more or less conflictual negotiation processes among the fishermen, either within a single gear group or among different gears'.

In multi-use situations where only fishers are involved, there are often informal regulations as to who may use certain locations at particular times. The means to avoiding conflicts is simply to segregate those using different kinds of gear into separate territories (Jentoft and Kristoffersen, 1989). Fishers from different villages on the Breton island of Houat have established a *modus vivendi* over all matters to do with the sharing of fishing grounds (Jorion, 1978). Two basic principles influence systems of informal territorialisation: first, fishing spots 'belong' to fishers from the nearest port; and, secondly, individual expertise gives the expert the right to use. In Spain, *cofradias* are responsible for territorial arrangements based on similar principles. But these principles are essentially ideal conditions which may not coincide with actual behaviour: internal conflicts continue to occur, while the encroachment of 'outsiders' into fishing territories traditionally claimed by certain categories of local fishermen pose increasing threats.

In some areas, institutions have been developed apparently around an 'ethic of fairness'. For example, the drawing of lots is intended to ensure that fishers take turns in exploiting good fishing spots; the same principle may also insist that they rotate spots after each set. The former system of allotment is practised and regulated by several *prud'homies* in the French Mediterranean (Dufour, 1996), while the rotation system is used in salmon fishing in an Irish village (Taylor, 1981, 1987), off the Greek island of Meganisi (Just, 1995) and in the swordfish fisheries in the Straits of Messina (Collet, 1996). Such institutions tend to work only when certain preconditions are met: first, the resources are considered communal property; second, the territories are respected by fishers from outside the area; and, third, that encroachment can be resisted. It should be understood, however, that territorialisation by fishermen usually has little to do with resource management, but rather with the creation of 'privileged space'. Such systems rarely find a place in modern, bureaucratic forms of resource management. *Cofradias* and *prud'homies* are involved in people management in the sense that they have powers to sanction those who breach the rules and regulations (Frangoudes, 1995). Although informal arrangements with respect to 'privileged space' exist in many parts of Europe and fishers will also attempt to monopolise the use of good fishing spots through the privatisation of knowledge as to their whereabouts, such locations are not formally owned in any way.

Placing limits on the numbers who can legally exploit renewable marine resources, through licensing, is a familiar mode of people management. Limited entry systems should avoid the problem of 'crowding'. In addition to effort controls involving gear restrictions or time limitations, limited entry is commonly used in single resource, single use management regimes as a means to avoid overexploitation. Such regimes are usually installed by external authorities. But governments or supranational institutions may crosscut systems of self-governance and rekindle conflicts between different user groups (Delbos and Prémel, 1996). When certain categories of fishers are excluded from access to a resource, while others are allowed entry, this may lead to strife within the fishing communities (van Ginkel, 1991).

A particularly difficult situation can arise over access to migratory species which may be caught before reaching inshore waters, creating problems for the less mobile small boat fishermen who are forced to stay within the coastal zone (Fingleton and McCann, 1996). Across a number of European countries there is a major dispute between commercial and recreational fishing interests over rights to fish the scarce wild salmon stocks. The netting of salmon has already been discontinued throughout most of the high seas fisheries in the North Atlantic, and is being phased out from inshore waters within the UK ostensibly to conserve the stocks - to the obvious concern of the commercial netsmen who feel that the inland sports fishing interests are being unfairly favoured. Both commercial and recreational fishermen jointly accuse the salmon farming industry of being responsible for the increase in sea lice infestation of the wild stocks of sea trout but these allegations remain unsubstantiated.

More generally, large scale offshore fishing competes with small scale inshore fishing for both resources and markets. In this respect, inshore fishers also face competition from sparetime or leisure fishers, who can catch considerable amounts of fish without being subject to the kinds of restrictions that apply to professional fishers and who may undermine quayside prices through direct sales to hotels and restaurants. Mondardini Morelli (1999) also points to intense conflicts between professional and leisure fishermen in Sardinia based on competition for marine resources and for port space.

2.2 Marine aquaculture

Aquaculture increasingly competes for space and resources with other users of the marine domain, among them inshore fishers. Fish farming and shellfish cultivation are expanding rapidly. Entrenched interests, such as traditional fishers who fear competition over fishing grounds or tourist agencies concerned over threats of 'aesthetic damage', often oppose the establishment of new aquaculture sites and attempt to maintain the *status quo* (Ridler, 1997). In Norway the expansion of marine aquaculture - mainly salmon farming - has already taken up much coastal space and with other marine species gradually being introduced into marine aquaculture, there will be an increasing need for multiple separation zones and other measures to avoid contamination and interaction effects (Sandberg, 1996). Along with the increasing crowding effect, the potential for conflicts between aquaculture and capture fisheries have grown. In some areas of Ireland, for example, fishers feel that they have had to sacrifice important fishing grounds to the interests of salmon farming, and there have been incidents where aquaculture cages have been sabotaged (Steins, 1998).

Inshore fishers are also naturally concerned that the quality of fishing grounds in the vicinity of fish and shellfish farming sites will deteriorate. More specifically they fear the introduction of non-native species, sedimentation of fishing grounds close to mussel farming areas, pollution by salmon farming through the accumulation of waste food, excreta and chemical residues and the concomitant decline in water quality.

2.3 The extraction of non-renewable resources

The discovery of commercial quantities of oil and natural gas in the North Sea in the 1960s affected the fishing industry in as much as competition for sea and harbour space, repair facilities and labour all increased. On the other hand, oil and gas development did offer alternative employment opportunities in remote coastal regions with poorly developed labour markets. Nonetheless, the overall view is that oil and gas exploration and development are unlikely to serve as a permanent replacement economy for most fishing communities, while they do have negative impacts on the fisheries. Fishers face a loss of access to fishing grounds, hazards to navigation, pollution from oil seepages, spills and blowouts; and damage to fishing gear as a result of debris dumped from rigs, supply boats and pipe laying barges. Fishing interests have generally held the weaker bargaining position in their negotiations with the oil and gas industry, though compensation schemes are available to cover part of the costs arising from lost fishing time and damage to gears (Grant, 1978; House, 1986).

Drilling rigs form physical obstacles and fishing is prohibited within a 500 m safety zone. Under-water oil and gas pipelines present similar obstacles to bottom trawling and will normally be avoided. But on the credit side, such structures simulate the functions of reefs, providing a refuge for fish populations.

In areas where land based activities associated with the oil and gas industry are concentrated, the inshore fisheries may suffer direct competition for labour, though it should be remembered that the oil and gas industry, once established, is capital intensive, employing relatively little labour much of which is in fact engaged in highly specialised technical or administrative work. Indeed many of the key workers are brought to the area from outside.

Possibly more serious from the fishing community's perspective is the tendency for higher prices to be charged for goods and services, including housing, in a boom economy.

Shell, sand and gravel extraction also poses threats to fishing activity, both directly and indirectly, albeit to a far more limited geographical extent than oil and gas exploitation, and over a much shorter time span - but often with more damaging consequences for the local fishing industry. Mineral extraction can disturb spawning and nursery grounds and bury active shellfish beds. Indirectly, it affects the ecological balance by causing longer term damage to the seabed and by increasing the turbidity of surface waters.

2.4 Non-extractive uses

Among the non-extractive uses of the inshore domain, exploitation of its amenity value would seem most likely to exert the greatest impact on fishing and fishing communities. Tourism has received a huge fillip as a consequence of increased leisure time, disposable personal incomes and improved transportation systems. Recent expansion of seaside tourism has spread outwards from existing coastal resorts, but over the past few decades has invaded previously untouched areas partly as a consequence of deliberate policies of local tourist boards and stimulated by the desire of holiday makers to spend their leisure time in 'unspoiled' areas of coast. The net result has been a gradual deterioration of the natural environment. Construction has altered the littoral sometimes causing considerable damage to marine life as for example when excessive sediment pollutes the coastal waters (Sibthorp, 1975). The development of coastal resorts and marinas can lead to pressures for space: fishers are sometimes relegated to peripheral locations, experience delays in moving in and out of crowded harbour areas and may experience growing opposition to their quayside activities (see Chapter 8). Cruise shipping, recreational boating, swimming, scuba diving and water skiing are among the new activities that can potentially interfere with inshore fishing and coastal aquaculture.

Although the rise of the coastal tourism and leisure industries can be observed throughout Europe, nowhere has its impact been greater than along the northern shores of the Mediterranean. By the mid-1970s, the Spanish Mediterranean coastline had been irreparably damaged by an uncontrolled spending spree: millions of foreign visitors invaded the coastal areas each year and it was reported that 'local fishermen have become barbers or waiters, wholly dependent on the tourist trade' (Sibthorp, 1975: p. 58). There are many examples where tourism interests, although relative newcomers to the scene, are better organised and much better represented in the decision making processes concerning coastal development and marine resource use than traditional user groups like fishers. On the other hand, tourism can provide an opportunity for fishers to earn additional incomes. Holiday making in coastal communities stimulates local demand for fish products. However, at the height of the tourism season, many local boat owners will make their craft available for hire for sightseeing or recreational fishing trips even when this coincides with the more important fishing season: tourists provide a more reliable source of income and require far less physical effort.

Not all new coastal development is associated with tourism. Coastal areas are preferred retirement locations; new port facilities are located away from congested established dockland areas, especially to accommodate the twentieth century's supertankers and bulk carriers; the new generation of nuclear and oil and gas fed power stations have been built in coastal locations, often in relatively remote sites; in some island fringed coastlines new road networks have necessitated the construction of bridges and causeways, while flood protection schemes have involved the building of massive dams, tidal barrages and thousands of kilometres of embankments. Many of these developments have affected the adjacent marine environment and impacted directly or indirectly on local inshore fisheries. Moreover, the increasing scale of seagoing navigation has extended the need to maintain shipping channels through frequent dredging.

2.5 Marine wildlife conservation

Inshore fisheries are subject to increasing interrogation from environmentalists regarding the impacts of their activities on marine habitats and ecosystems (see Chapter 14). In recent times, environmental organisations have become identified as legitimate stakeholders; in some instances they clearly strive for hegemony in the decision making process concerning the inshore domain and are reluctant to compromise their objectives (Chapter 5). Most environmental organisations lay stress on the natural values of the seascape and some are apt to favour limited use or even non-use of renewable resources through their advocacy of marine protected areas (MPAs), though in most instances such designations seek to restrict rather than exclude fishing activity. But the very suggestion of MPAs is liable to force 'a polarisation of interests rather than a dialogue between interested parties ... creating instances of tension and conflict' (Mondardini Morelli, 1998: p. 188). To an extent, some conservationists have sought to turn fish and shellfish into products of zero exchange value and, in so doing, have attempted to remove them from the commodity path. As protectors of the marine domain, conservation organisations enjoy widespread popular support and have achieved considerable political influence, in marked contrast to inshore fishing interests.

2.6 Waste disposal and pollution: abuse of the marine environment

Macrostructural social changes such as population growth, industrialisation and urbanisation have led directly to increasing discharges of effluents and the disposal of polluting agents in coastal waters. Most marine pollution originates on land and is channelled to the sea through the river systems and major estuaries. As a result coastal waters and semi-enclosed seas tend to be much dirtier than the open oceans. Urban, industrial, agricultural and, in some cases, mining waste thus enters the marine environment putting at risk the productivity of the inshore ecosystems and the viability of fish and shellfish populations. Pollution of the marine environment is typically a transboundary problem which has to be tackled internationally and within the context of the regional seas - a point readily acknowledged by international conventions and regional organisations like the North Sea ministerial conference (Svelle *et al.*, 1997).

Throughout much of the industrial age the seas have been consciously or unconsciously regarded as a cheap dumping site for various kinds of waste material: dispersion and dilution of potentially harmful materials were considered the natural functions of the oceanic systems. The release of untreated sewage into coastal waters patently conflicts with their natural use as spawning and nursery grounds and their applied use for fish farming and shellfish cultivation. Sewage can be directly toxic to living organisms and cause oxygen depletion through nutrient enrichment. In general, the smaller the area of water and the weaker its circulation, the worse will be the effects of sewage discharges. From time to time governments must act to close down shellfish production in waters where there is a risk it becoming so tainted or contaminated that it is unfit for human consumption. The release of nutrients into coastal waters through urban sewage disposal or agricultural land run-off can also lead to blooms of toxic plankton and algae posing the risk of substantial mortalities among fish and shellfish populations in inshore waters. Mortality and physical abnormalities can also result from high concentrations of heavy metals and PCBs resulting from the discharge of industrial or mining waste water.

One of the most dramatic forms of impact upon the marine ecosystems of inshore waters and their associated fisheries results from catastrophic episodes of oil pollution caused either through accidents or through purposive acts of cleaning out oil tanks at sea. Shipwrecks, groundings and blowouts on offshore oil rigs account for the worst oil related environmental disasters. As to the severity of the incident much will depend on whether the spillage is of heavy or light oils, but in any case oil is relatively difficult to disperse or contain in water. Incidents with oil tankers like the Torrey Canyon, Amoco Cadiz, Braer, Sea Empress and, recently, the Pallas, Erika and the chemical tanker Ievoli Sun have had devastating effects on the inshore marine environment, marine wildlife and fishing activity. Major blowouts like the one in the Ekofisk field in 1977 and the Piper Alpha disaster in 1988 also released huge amounts of oil into the sea but their effects on the inshore marine environment were to some extent mitigated by the fact that the incidents occurred in the open sea rather than close inshore.

More oil enters the marine environment through discharges and industrial effluents than through accidents. Oil tankers deliberately discharge significant amounts of oil through cleaning their tanks before taking on new cargoes and many vessels also empty oil contaminated ballast and bilge water while at sea, usually close to port. The normal operation of coastal refineries and industrial plants adjacent to estuaries can lead to oil leakage or seepage into the tidal waters. Oil pollution, even in small doses, taints the flesh of fish and especially shellfish making it unmarketable; more severe contamination can also have serious consequences for the reproduction success of some marine species (Sibthorp, 1975).

3. OPPORTUNITIES FOR MULTI-USE MANAGEMENT

Several international treaties and conventions relating to the prevention of marine pollution are already in force and in recent years regulations have been progressively tightened. Indeed, one of the success stories of the EU's environmental policy has been the significant improvements in water quality in both freshwater and marine environments, but there is still a good deal to be done. In a recent policy document, the International Maritime Organisation (1998) observes that measures which have already been adopted should be implemented as rapidly and as effectively as possible and that governments should act to enforce the relevant international conventions. While it is clear that environmental problems like marine pollution can only be tackled effectively on a global scale and through international collaboration, the most viable means of tackling specific multi-use problems in inshore waters occur at the local level.

Coastal planning involving all stakeholders and integrating all the options is the preferred way of managing competing uses and interests. An integrated approach would require the weighing of costs and benefits of each conflicting activity (Ridler, 1997). This was recognised, as early as 1975, in respect to the North Sea.

> 'To make the best use of it in the aesthetic and recreational sense as well as commercially, rational management is vital to ensure both that each activity is conducted at such a rate and in such a way as to serve the best long term interest, and that the prosecution of one activity is harmonised, so far as possible, with that of another'.
>
> (Clark, 1975: p. ix)

The aim of multiple use management should be the optimal exploitation of resources, with the various interacting activities managed in such a manner as to achieve the greatest overall benefit. It is beyond the scope of this chapter to calculate the relative importance of widely disparate activities carried out in the inshore zone. Instead, we are content to outline the basic means of negotiation concerning access to, allocation of and control over marine resources on behalf of different user groups. In order to achieve legitimacy in the view of different users, it will be necessary to adopt specific management measures which meet with widespread approval and are ultimately enforceable. The means to achieve this vital consensus lies in the concept of co-management which aims to forge closer links between all relevant stakeholders and the central and regional authorities with appropriate legislative powers. A basic construct in co-management is the establishment of 'user platforms' where all interest groups can meet to debate the issues and tease out a consensus view.

Co-management organisations are negotiating and/or decision making bodies, formed on a voluntary or statutory basis, comprising different stakeholders who work collectively towards an understanding of the resource issues, cooperate in solving social dilemmas associated with collective resource use, and undertake joint action in respect of perceived problems. The need for such bodies may arise spontaneously when stakeholders experience the negative impacts of their own and other users' exploitation of the natural resource and become aware that concerted action is needed to resolve undesirable outcomes. Experience from around the world - and not only in relation to fisheries - suggests that the establishment and facilitation of organisations that bring together various interest groups has considerable potential for solving natural resource management problems through collective action and joint learning based on the accommodation of different and often conflicting objectives (Röling and Wagemakers, 1998).

Although *cofradias* and *prud'homies* can be considered as the predecessors of modern co-management organisations for the collective management of a single use, what is 'new' about the idea of 'user platforms' is their potential to facilitate collective action among separate user groups in complex, multiple use situations without necessarily involving the central or regional bureaucracy. For instance, in Killary Harbour in Ireland, the development of mussel farming required the renegotiation of resource use which traditionally focused on salmon, lobster and crab fishing. When it was discovered that the estuary was an excellent mussel spawning area, local community members established a mussel farming cooperative for local mussel farmers only, in order to avoid the colonisation of the resource by 'outsiders'. At the time of the cooperative's inception, local fishermen were concerned that the estuary would be filled with mussel structures, preventing access to their fishing grounds. The cooperative, aware of the fishermen's fears, agreed to keep the fairway and fishing grounds free from rafts and longlines, even though these areas were designated for mussel production. The cooperative, assisted by a community development project in the area, was able to convince the government to support its policies.

At the regional level, co-management can also involve the elaboration of integrated coastal zone management strategies. In the case of the Bantry Bay Project, in 1997 a partnership comprising Cork County Council, the Coastal Resources Centre based in University College Cork and the Nautical Enterprise Centre secured funding under the EU LIFE programme for a project to develop sustainable ways of managing the land and water resources of the bay. Bantry Bay in south west Ireland is one of the most important tourism and nature conservation areas in the country; it also harbours Ireland's second largest fishing port (Castletownbere) and has become a prime aquaculture production area. Lack of coordination between different users had resulted in a number of multi-use problems. In particular, there was growing conflict between mussel cultivation, on the one hand, and tourism and fishing interests, on the other. The Project brought in external experts specialising in conflict resolution to undertake a consensus building exercise: the Centre for Dispute Resolution in Dublin has worked closely with the various stakeholders in establishing a process which involves:

(i) working at understanding each other;
(ii) looking for ways of satisfying each other's needs, interests and concerns;
(iii) building on areas of common ground while trying to reduce differences;
(iv) exploring the scope for mutual gain;
(v) confronting the problems not each other; and
(vi) investing in relationships for the future (Bantry Bay Charter, 1997).

Consensus building is regarded as a crucial element in the overall project (Centre for Dispute Resolution, 1998).

The strategy for integrating shellfisheries and nature conservation in the Dutch Wadden Sea, outlined in Chapter 5, offers another example of the co-management approach - this time developed at the national level - based on creating a platform for resolving key disputes between the fishing and conservation interests. In this case the emergence of difficulties at the start of the second management plan simply underlines the importance of building confidence, mutual trust and genuine consensus through the user platform idea.

The formation and facilitation of user platforms for resource use negotiation therefore offers a promising but certainly not unflawed mechanism for consensus building over responsible behaviour in exploiting inshore waters. Several issues are of importance in attaining this goal. First, the scale of the platform must correspond with the scale of the ecosystem to be managed and with existing economic and administrative institutions. Second, it is important that all relevant stakeholders are correctly identified at the outset in order to examine all sources of pressure on the resource, the full range of vested interests and the relative stakes held by each user group. The alienation of certain user groups through their exclusion from the process risks the erosion of legitimacy for any decisions adopted (Symes, 1996). Third, management of the process must focus on social processes and collective learning about the situation to be tackled. Finally, the role of external agents must be addressed with great sensitivity. While they may be essential for facilitating and supporting actions by local groups and help to reduce transaction costs, the over-involvement of external agencies can undermine the basis of cooperation and the sense of ownership and responsibility for the project. Much is to be gained by seeking to give local associations a more permanent footing and to bestow on them certain management responsibilities.

4. CONCLUSION

Broadly there are two contrasting approaches to resolving multi-use problems: legislating at national or international level for more efficient regulatory frameworks, on the one hand, and seeking compliance through legitimacy negotiated at the grass roots level, where mutual respect and true consensus are the precursors to the enactment of new regulations, on the other. Both are needed to tackle the kinds of problems facing inshore fisheries outlined above. Problems whose origins lie outside the local area - as in the case of the oil and gas industry, urban development and pollution - must be addressed through national and international legislation; those which are internal to the area - including competition within fisheries and between fisheries, aquaculture, tourism and marine conservation should be subject to local arbitration within some form of overarching strategy for the integrated management of the inshore zone.

The creation of formal links and channels of communication between all parties is a first requirement for a properly integrated approach. The presence of multiple, but unconnected, user groups, external agencies and authorities will in fact hinder integrated management and increase the likelihood of multi-use conflict and irresponsible behaviour. Formalised links between the parties involved will facilitate negotiation over the formulation of management objectives, strategies, action plans, the division of responsibilities and the elaboration of monitoring procedures. In situations where such links have been established it is important that they are not restricted simply to the time when policies are being formulated and when the integration of particular interests is initially at stake but persist throughout the ensuing period. The challenge of multi-use management is to integrate disparate interests so as to achieve socio-cultural acceptability, economic feasibility and environmental sustainability (Glaeser and Píriz, 1996). Ongoing dialogue is required to create mutual understanding of each other's perspectives, interests and concerns in order to avoid an 'us' and 'them' situation arising when negotiation over policy is taking place and to prevent stronger sectoral interests overruling the weaker ones. Opportunities for dialogue should persist well into the implementation stage to ensure that the distributive effects of the policy are as anticipated and agreed during its formulation.

It is important to identify the vested interests and relative stakes held by different user groups and to understand the ways in which people, resources and institutions in multi-use contexts are embedded within webs of economic, social and cultural relations at various levels of integration. Marine resource use and marine environmental issues transcend nation state boundaries and a meld of local, national and international actors have identified themselves as principal stakeholders with regard to use rights over marine space and access to resources. It is precisely this fact which can lead to a clash of interests between the various claimants. In many cases, there will be power hierarchies or structural asymmetries of stakeholder groups and individuals, whose socio-economic and cultural statuses vary. In negotiations, some may be included, others excluded. The problem of participation, representation and cooperation in decision making has important consequences for the possibility of disputes and dispute regulation. The interests of local fishers often clash with state imposed regulations concerning the use of the inshore zone. In the past, governments have tended to favour the economically stronger sectors like aquaculture, tourism or oil exploration over the economically weaker inshore fishing interests. Fishers thus risk being alienated from the resources they traditionally exploit and that sense of alienation may well undermine the legitimacy of measures adopted. Management needs to take account of the socio-economic and cultural contexts of the different types of user.

It is equally important for an integrated management approach to decentralise at least part of the decision making structures. Multi-use problems can best be identified and tackled at the local level. Even though some sectoral interests may well exceed this level, management strategies need to be locally compatible because compliance with rules and regulations largely depends on public support and local participation. There is evidence that under centralised and non-participatory approaches, sectoral development and the non-integration of different interests can lead to severe conflicts. In a decentralised system, local authorities must be given powers to react immediately to emerging problems without the need first to consult the central state, a time consuming process which hinders flexible management. Further, local agreements should not be capable of being challenged by other agencies. User participation, balance of power and consensus building are not goals in themselves, they are the means to an end. Though the importance of these social processes can hardly be underestimated, collective learning about the marine ecosystem, its complex dynamics and varied functions is an equally important component in balancing the interests of multiple users and achieving harmonious resource use.

5. REFERENCES

Acheson, J.M. (1988) Patterns of gear changes in the Maine fishing industry. Some implications for management, *Maritime Anthropological Studies*, **1** (1), 49-65.

Bantry Bay Charter (1997) *Bantry Bay Charter, Issue 1,* November,1997.

Centre for Dispute Resolution (1998) *Progress Report 1 on the Consensus Building Programme for Bantry Bay Project*, Centre for Dispute Resolution, Dublin.

Clark, R.B. (1975) Introduction, in M.M. Sibthorp (ed.), *The North Sea. Challenge and Opportunity*, Europa Publications, London, pp. ix-xiii.

Collet, S. (1996) Leviathan management or customary administration: the search for new institutional arrangements, in K. Crean and D. Symes (eds.), *Fisheries Management in Crisis*, Blackwell Science, Oxford, pp. 99-109.

Delbos, G. and Prémel, G. (1996) The Breton fishing crisis of the 1990s: local society in the throes of enforced change, in K. Crean and D. Symes (eds.), *Fisheries Management in Crisis*, Blackwell Science, Oxford, pp. 129-140.

Dufour, A-H. (1996) Management and practice in the small-scale inshore fisheries of the French Mediterranean, in K. Crean and D. Symes (eds.), *Fisheries Management in Crisis*, Blackwell Science, Oxford, pp. 110-115.

Edwards, V.M. and Steins, N.A. (1998) Developing an analytical framework for multiple-use commons, *Journal of Theoretical Politics*, **10** (3), 347-383.

Farrugio, H. and Le Corre, G. (1991) Recherche halieutique et pêches artisanales interactives en Méditerranée française, in J.R. Durand, J. Lemoalle and J. Weber (eds.), *La Recherche Face à la Pêche Artisanale*, Editions de l'ORSTOM, Paris, pp. 349-355.

Fingleton, P. and McCann, S. (1996) The use and regulation of coastal resources. Traditional structures and modern national and EU overlays in Western Ireland, *MARE Working Paper* No. 7, Aarhus University, Aarhus.

Frangoudes, K. (1995) Changing fisheries management institutions in the French Mediterranean, *MARE Working Paper* No. 4, Aarhus University, Aarhus.

Frangoudes, K. (1996) Fishermen households and fishing communities in Greece: a case study of Nea Michaniona, in K. Crean and D. Symes (eds.), *Fisheries Management in Crisis*, Blackwell Science, Oxford, pp. 119-128.

Glaeser, B. and Píriz, L. (1996) Sustainable coastal zone management. Competing interests and conflict resolution in Sweden. Paper presented at *On Northern Shores and Islands, International Nordic Conference in Human Ecology*, Strömstad (Sweden), 16-18 August 1996.

Grant, J.P. (1978) The conflict between the fishing and oil industries in the North Sea. A case study, *Ocean Management*, **4**, 137-149.

House, J.D. (ed.) (1986) *Fish Versus Oil: Resources and Rural Development in North Atlantic Societies*, ISER, Memorial University of Newfoundland, St. John's.

International Maritime Organisation (1998) *Preventing Marine Pollution*, 19 March 1998, IMO, London.

Jentoft, S. and Kristoffersen, T. (1989) Fishermen's co-management: the case of the Lofoten fishery, *Human Organization*, **48** (4), 355-365.

Jorion, P. (1978) Marks and rabbit furs: location and sharing of grounds in coastal fishing, *Peasant Studies*, **7** (2), 86-100.

Just, R. (1995) The Meganisiot fishing system: equity and self-interest, in S. Damianakos (ed.), *Brothers and Others. Essays in Honour of John Peristiany*, EKKE, Athens, pp. 241-252.

Mondardini Morelli, G. (1998) Traditional use rights in the small-scale fisheries and marine protected areas between Sardinia and Corsica, in D. Symes (ed.), *Property Rights and Regulatory Systems in Fisheries*, Blackwell Science, Oxford, pp. 188-200.

Mondardini Morelli, G. (1999) Fishery and tourism in the Mediterranean: the case of Sardinia, in D. Symes (ed.), *Europe's Southern Waters: Management Issues and Practice*, Blackwell Science, Oxford, pp. 78-87.

Ridler, N.B. (1997) Rural development in the context of conflictual resource use, *Journal of Rural Studies*, **13** (1), 65-73.

Röling, N.G. and Wagemakers, M.A.E. (1998) *Facilitating Sustainable Agriculture: Participatory Learning and Adaptive Management in Times of Environmental Uncertainty*, Cambridge University Press, Cambridge.

Sandberg, A. (1996) Use and protection of coastal resources, *MARE Working Paper* No. 7, Aarhus University, Aarhus.

Sibthorp, M.M. (ed.) (1975) *The North Sea. Challenge and Opportunity*, Europa Publications, London.

Steins, N.A. (1998) We have to keep the foreigners out of our bay: top-down regulations and the strategic response of Irish fishermen, in D. Symes (ed.), *Property Rights and Regulatory Systems in Fisheries*, Blackwell Science, Oxford, pp. 216-229.

Steins, N.A. and Edwards, V.E. (1999) Platforms for collective action in multiple-use common pool resources, *Agriculture and Human Values*, **16** (3), 241-255.

Svelle, M., Aarefjord, H., Heir, H.T. and Øverland, S. (1997) *Assessment Report on Fisheries and Fisheries related Species and Habitats Issues*, Ministry of the Environment, Oslo.

Symes, D. (1996) Co-governance in marine and coastal fisheries, in *Proceedings of the Seminar on Co-Management 'Co-operation in Management to the North Sea and Wadden Sea'*,

Ministry of Agriculture, Nature Conservation and Fisheries, Fisheries Directorate, The Hague.
Taylor, L.J. (1981) 'Man the fisher': salmon farming and the expression of community in an Irish settlement, *American Ethnologist*, **8** (4), 774-788.
Taylor, L.J. (1987) 'The river would run red with blood': community and common property in an Irish fishing settlement, in B.J. McCay and J.M. Acheson (eds.), *The Question of the Commons. The Culture and Ecology of Communal Resources*, The University of Arizona Press, Tucson, pp. 290-307.
van Ginkel, R. (1991) The sea of bitterness: political process and ideology in a Dutch maritime community, *Man* N.S., **26** (4), 691-707.
van Hoorn, H., Peet, G. and Tiel, M. (eds.) (1980) *The North Sea and its Environment. Uses and Conflicts*, North Sea Working Group, Amsterdam.

Chapter 16

INTEGRATED COASTAL ZONE MANAGEMENT: THE SWEDISH EXAMPLE

Laura Píriz

1. INTRODUCTION[1]

Nature in general, and the coast in particular, occupy a prominent role in the life of the Swedish population and the concepts of welfare and standard of living are very much related to access to the coast for recreation and leisure purposes. Approximately half of Sweden's population resides permanently within a 30 km wide coastal zone (Bernes and Grundsten, 1991). In addition, a third of the 600,000 properties registered as 'summer houses' (Statistics Sweden, 1998) are situated within 1.5 km of the coast or lake shorelines (Segrell, 1995).

The previous chapter has highlighted how throughout Europe the coastal zone, once the exclusive domain of fishers and sailors, is now increasingly being used for a range of purposes other than fishing and navigation. Historically, the combination of activities such as fishing and farming (fisher-peasants) has had enormous significance for the economic and social development of the Swedish coastal margin; the importance of the traditional pluriactive economy can still be seen reflected in the coastal settlement pattern and in the rural landscapes, though now considerably altered under the influence of the 'third wave' of recreational development. Maritime trade and the shipping industry, along with defence, were also significant formative influences helping to identify important nodes of additional activities within the settlement pattern. Much of the urban and rural architecture dating from before the mid-twentieth century mirrors the

[1] The chapter includes information prepared within the framework of the Swedish research programme on Sustainable Coastal Zone Management (SUCOZOMA) funded by the Foundation for Strategic Environmental Research, MISTRA.

D. Symes and J. Phillipson (eds.), Inshore Fisheries Management, 275–293.

importance of these traditional activities and the symbiotic cultural relationship between land and sea.

Today these activities, like the coastal fisheries themselves, are confronting a period of recession, at least in terms of employment. Since the 1950s, and especially in peripheral areas, fisheries dependent communities in Sweden have undergone a period of drastic population decline and decreased dependency on coastal fisheries (Píriz, 2000). In general, resource management institutions, particularly those focusing on a particular sector of the economy, have been remodelled to meet the needs of centralised and capital intensive fisheries. The difficulties confronted by the localised coastal fishery and their resolution are not only related to the processes and decisions occurring within the sector but also to the many interactions with other sectors of activity and a much wider range of policies.

The diversification of the coastal economy from a simple commodity production based on farming, fishing and forestry (see Chapter 11) has evolved over the past two hundred years. First, the coastal zone became the location for industrial, land based activities that were not directly dependent on marine living resources for their existence but where a coastal location proved to be economically efficient through savings in costs of assembling the basic factors of production, distributing the finished goods or disposing of waste materials. A more diffuse and somewhat later development was the use of the coastal zone for permanent or seasonal residence: large tracts of Sweden's coastline became extensions of the living space for the expanding urban population. With industrialisation and the increasing need for nature based leisure activities mainly for people from the urban areas, the coastal zone has evolved over time into an area where the values and interests of non-residents frequently predominate. In such a situation there is always a risk that what localised, small scale coastal fishers need to structure their profession becomes lost in intra- and inter-sectoral trade-offs.

Outdoor recreation and tourism including recreational fisheries and seal safaris, together with newly evolving plans for sea based, wind power parks and new ideas for expanding the cultivation of mussels are all activities that, developed in relation to strict environmental criteria, have a potential for expansion in coastal Sweden. The same is, of course, true for actions related to nature conservation such as the designation of marine protected areas (MPAs). Recently, reference to the precautionary principle and the pre-requisite of maintaining biological diversity and ecosystem service functions in natural resource management has been impacting on nature conservation policies in Sweden. These developments serve to make the valorisation of natural resources - already an ongoing process - much more strongly evident in coastal management. In transforming the coast into an area where alternative solutions to land based problems are to be located, old and new interests overlap and potentially conflict.

The outcomes of the implementation of both sectoral and wider regional policies do not necessarily meet the specific needs of local coastal fishers in the archipelago, even when paradoxically a local coastal fishery is expected to play an important role in sustaining livelihood strategies in accordance with instructions mandated by the Swedish parliament for 'ecological sustainable development' (Riksdagen, 1996) and 'a living archipelago' (Miljödepartementet, 2000). According to these concepts, the maintenance of population levels and the socio-economic stability of settlements in the Swedish archipelago are seen as the two main requirements for sustainable development in which the coastal fisheries are expected to play a strong role.

To date, a comprehensive account of the legislation and practice dealing with the management of the coast in Sweden is lacking. In 1995 the Swedish Environmental Protection Agency issued a report purporting to describe the Swedish model for coastal zone management (Ackefors and Grip, 1995) and in the same year Swedmar published a concise description focusing on the municipal planning process (Johansson, 1995). Surprisingly in neither report is mention made of informal procedures such as the operational and collective rules governing the activities of coastal fishers or of the large numbers of ongoing initiatives and projects and the application of structural funds in the coastal zone.

Sweden has a long tradition of regulating and planning the use of land. Experience of shoreline management is rather more recent, though still often referred to in international literature as a good exemplar of integrated coastal zone management. The Swedish legal and regulatory framework that, in one way or another, addresses or influences the use of coastal resources is quite robust, affecting the different sectors and levels of decision making in the administration quite profoundly. To address the whole system goes well beyond the scope of this chapter. As the systems of coastal fisheries management have already been outlined in Chapter 3, the present aim is to examine how far other coastal management instruments and procedures impact on the needs of the coastal fishers. Focus is maintained on the legal framework and formal management instruments, especially those which address competing interests and the vertical and horizontal integration of management in a multiple use context. In the second part of the chapter, the circumstances in which these management instruments may be perceived as threats by coastal fishers are examined.

2. TOOLS FOR THE INTEGRATION OF COASTAL MANAGEMENT IN SWEDEN

2.1 The Natural Resources Act 1987: a legal precedent for 'ecological sustainable development'?

Sweden's environmental legislation was reformed in the closing years of the 1990s with the pre-existing principal acts consolidated into an Environmental Code which came into force in January 1999. The effect has been to harmonise environmental law and to develop instruments to drive environmental policy forward. In the new Code, roles have been clarified, new principles incorporated and rules and instruments dealing with species, biotopes and area protection brought together. Moreover the function of the Environmental Courts has been strengthened and the setting of environmental standards introduced.

A central point of reference for the Swedish resource management system, and one which relates to the concept of 'ecological sustainable development' referred to above, is the *Natural Resources (Management) Act 1987*, an umbrella act covering all legislation relating to the regulation of land, water and the physical environment. The Act established that 'land, water and the physical environment are to be used in such a way that they ensure long term good husbandry from an ecological, social and socio-economic point of view' (Svensk Författningssamling, 1987: p. 12, author's translation). Subsequently the Environmental Code has confirmed that

> 'Land and water areas shall be used for purposes for which they are best suited ... Priority shall be given to use that promotes good management from the point of view of public interest ... Land and water areas that are particularly vulnerable from an ecological point of view shall, as far as possible, be protected against actions that may damage the natural environment ... [and] areas that are important for ... commercial fishing or aquaculture shall, as far as possible, be protected against actions that may significantly interfere with the operations of these industries'
>
> (Svensk Författningssamling Ds, 1999: iv)

The prioritisation of the natural environment and the protection of natural resource based industries is thus made clear.

The original 1987 Act deliberately did not adopt a bio-centric approach to resource management but rather an anthropocentric and utilitarian position. The notion of sustainable development is included and, through the way in which the Act is constructed, it internalised the dilemma of development

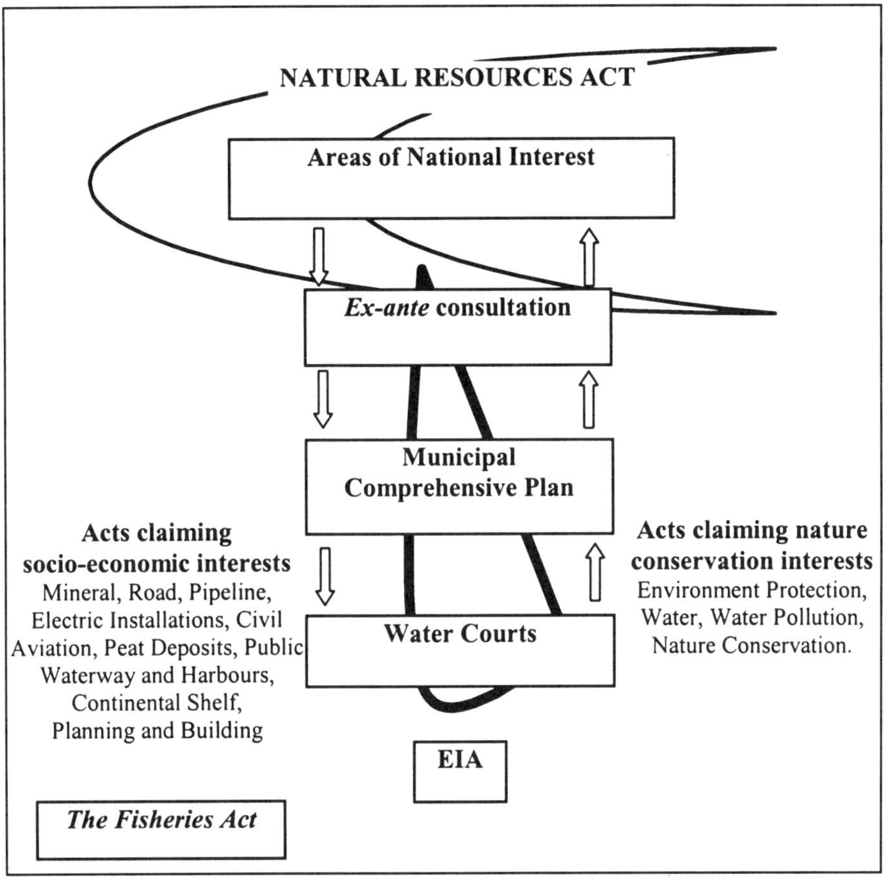

Figure 16.1 Sweden: The Natural Resources Act, 1987

versus conservation. Figure 16.1 illustrates the two aspects of the Act subsequently embodied in the Environmental Code: on the left is a series of Acts which may be said to establish socio-economic grounded rights to nature; on the other are those Acts which essentially defend the rights to both protect and enjoy the direct benefits of nature. The same is broadly true for the more recent Environmental Code. Overall, however, the Code adopts a less utilitarian and development oriented approach with the adoption of additional goals relating to human and environmental health, natural and cultural environments and biological diversity, *inter alia*. The intention of the 1987 Act was to provide a common basis for decisions against which other more specific issues have to be scrutinised by the County Administration Board on behalf of the central government.

The Natural Resources Act also included both general and specific management provisions for sectoral development, on the one hand, and for the development of geographical areas delimited on the basis of the natural and cultural features, on the other. These prioritised sectors and areas were to be designated by the central authorities as Areas of National Interest. The Act further stipulated that certain areas shall be used for certain economic activities also of 'national interest': these include agriculture and forestry; fishing; the extraction of raw materials such as minerals, peat or sand; industrial production; energy supply and communications; water supply and waste treatment; and national defence. But the Act also stipulated that such developments are only to be permitted providing they have no detrimental effects on: large, unexploited areas and ecologically sensitive areas; natural and cultural environments of 'national interest' for conservation and research; and areas set aside for outdoor and other recreational activities.

Provisions in the Act were not, however, intended to hinder the further development of existing settlements, local industry or national defence installations. The Act also reserved areas where recreation and tourism are to be given priority, many of which are, of course, located in coastal and archipelago areas. Here the central administration has the authority to prevent the introduction of certain types of industrial activity.

The formal consultation process (both vertical and sectoral), the Water Courts - now known as the Environmental Courts - and environmental impact assessments (EIAs) are all integral to the implementation of the Natural Resources Act and subsequently the Environmental Code. While these instruments are used mainly to assess environmental compatibility, the Comprehensive Municipal Plans attempt to direct local development, assess sectoral compatibility and expose intersectoral externalities and have not therefore been incorporated within the new Environmental Code. Although in Sweden there is a long tradition of comprehensive land use planning and consultation, the use of environmental impact assessment is more recent and as yet less well developed. Indeed the implementation of environmental impact assessments has been strongly criticised (Westerlund, 1992), largely but not solely on account of poor praxis and a lack of technical guidelines.

The formal rules for fisheries management, developed within the framework of the Fisheries Act 1993 and additionally by means of periodic fisheries regulations, were not subordinated to the Natural Resources Act and they too have not been included in the Code. Regulations formulated within the framework of the Fisheries Act are sector grounded and as such not intended to be integrative nor to concern themselves with issues other than fisheries conservation. Decisions taken within the frame of the Fisheries Act will, of course, have socio-economic and distributive effects. But, even in this context the Fisheries Act is not considered by most officials at the

National Board of Fisheries in Sweden as providing any guidance in relation to differential access to the resource.

Seawards the remit of the Natural Resources Act extends 3 nautical miles from the baseline, an area that is essential to the coastal fisheries. Even though the Fisheries Act does not directly come under the auspices of the Natural Resources Act, coastal fisheries inevitably receive special consideration. Similarly, areas intended for outdoor recreation and land and water of significance to professional fisheries and reindeer herding have to be identified by the central administration and given certain priority in the planning process. Taking these principles into consideration, together with the notion of ecologically sensitive areas of national interest, all shallow coastal waters up to 6 m in depth in unexploited or lightly exploited areas are to be protected. Such areas of national interest are subsequently given special consideration in the application of other planning instruments such as environmental impact assessments and the Environmental Courts.

The Natural Resources Act also defines certain activities which require special government permission for their establishment. Among these are heavy chemical industries, oil refineries and sea platforms for wind power generation. Prior to any major development decision, such as the expansion of wind power, the government will normally commission an appraisal. Through this and other mechanisms, the sectoral, regional and local organisations representing different stakeholder interests are consulted. Nonetheless the Natural Resources Act has been criticised both for its vague formulation and its subordination of local interests in favour of national interests. This conditioning is often manifested when Comprehensive Municipal Plans are prepared or the development of new industries considered.

The County and Municipal administrations play important roles in relation to the implementation and enforcement of the Natural Resources Act and subsequently the Environmental Code. The same is also true when it comes to *ad hoc* projects aimed at improving the environmental situation along the coast, even when the initiative was originally triggered at a higher level as the result, for example, of a parliamentary decision, EU directive or international convention. Currently there are several important initiatives which will impact directly or indirectly on the coastal fisheries: these include the implementation of natural resource and environmental management plans for the archipelagos (Miljödepartementet, 2000); the determination of water quality objectives; the integrated management of special areas such as Gullmar fjord; and, last but not least, the designation of Natura 2000 sites and negotiations in respect of marine protected areas.

The *Shore Protection Act 1964* and the concept of *allemansrätt* reinforce the special consideration given to outdoor public recreation by the Natural

Resources Act. The implementation of the Shore Protection Act is intended to serve the recreational needs of the Swedish population in general by securing access to sites for bathing and outdoor recreation and preventing the building of summer houses within 100-300 m of the shore. The application of these provisions has been devolved to the municipalities, though grants for exemptions have to be approved by the County administration. Segrell (1995) in an in-depth analysis of the history and significance of the Shore Protection Act asserted that NGOs with interests in the protection of nature or the promotion of outdoor recreation have generally been critical of the decentralised application of the Act. 'Above all, the Society for the Conservation of Nature has strongly opposed almost all proposals that have aimed at increasing the municipalities' influence' (Segrell, 1995: p. 189).

Over time a new interpretation of the Shore Protection Act has evolved that acknowledges the primary function of protecting nature and grants the option to the counties to extend the protected zone from 100 to 300 m from the shoreline. This extension has been exploited particularly in coastal areas classified as unexploited or with special natural and cultural value. Coastal buildings such as workshops needed for the daily operation of the professional fishers are exempted from this ban on development.

Allemansrätt is the practice which permits the public free access to nature, land and water even on private property provided one behaves in a socially acceptable manner. In Sweden *allemansrätt* is not applicable to fishing. Nevertheless, with the development of recreational fisheries and the commercialisation of fishing based tourism, individuals and companies have sought to claim access rights based on *allemansrätt*.

2.2 The Environmental (Water) Courts

Under the provisions of the Environmental Code, permits are required for all kinds of construction work which might modify the local, regional or national aquatic environments. These permits are issued by the Environmental Courts, previously known as the Water Courts. They were originally developed to serve the needs of hydroelectricity power plants in riverine locations. Today, examples of applications for permits brought before the Environmental Courts involve pipelines, cables, marinas, quarries, road constructions and so forth. The main task of the Court is to determine the conditions under which these constructions can best be undertaken with minimal disruption to the environment or to established economic activities. The National Board of Fisheries may be asked to represent the fisheries sector in this instance. Programmes for monitoring the outcomes of the development in relation to fish and fisheries and compensation for any damage caused are the most common conditions advocated by the Board. Compensation is not necessarily used to mitigate the effects of damage locally but rather to enhance fish stocks elsewhere by ranching or restoring biotopes. As a rule, when a major development project is planned by a public authority at national, regional or local level or by an established private enterprise, a process of informal consultation precedes the application to the Environmental Court. During such a process, where appropriate, the National Board of Fisheries can seek to influence the design and site selection for the development. The Board's officers will collect information from the fishermen's organisations and specialist research centres and assess its relevance (Johansson, 1998).

Among the most recent proposals brought to the attention of the fishing industry and the media are for the construction of a new hydroelectric power plant in the north of the country, sea based wind power parks on the west coast and submarine cables to transport electricity to and from Europe in the south. In all cases these developments are related to the search for viable alternatives to nuclear power generation.

2.3 Sectoral coordination

At present the coordination between different sectoral interests over the use and management of coastal and marine resources in Sweden is incorporated in several different processes operating at different levels. With increasing emphasis given to environmental and nature conservation policies, the integrating role of agencies representing these interests has grown considerably. In addition, each sectoral authority, including the National Board of Fisheries, has been given environmental responsibilities and instructed to set environmental goals and monitor progress by means of measurable indicators. Voluntary cooperation between authorities along both horizontal and vertical axes is now very common and is based on dialogue, consultation and joint working groups. It involves the exchange of memoranda, assessments and draft decisions for comment on an *ex-ante* basis. Participation in the consultation process normally builds on the representation of particular interests. In this way those who represent particular stakeholdings can elaborate their position and present arguments either in favour of or against specific developments.

Sweden has also experimented with the idea of coordinating secretariats. In the 1970s the Swedish Marine Resource Commission (*Delegationen för Samordning av Havsresursverksamheten*) was created for the coordination and planning of the use of marine resources. At that time the development of new marine based industrial activities, including aquaculture, oil exploration,

Box 16.1 The Marine Resource Commission's proposals concerning the utilisation of living marine resources in Swedish waters

- to pay more attention to the significance of the coastal fishery;
- to identify at national and regional levels the inputs needed to guide fisheries development;
- to distribute more effectively the support provided on a regional and sectoral basis;
- to analyse the possibility of extending trawling restrictions in areas where conflicts occur between coastal and offshore fishers;
- to assess the effects of demersal trawling on spawning areas and bottom fauna; and
- to enhance fisheries research, particularly at the National Board of Fisheries, concerning environmental impacts on fish and fisheries and multi-species models of stock assessment.

Source: Anon, 1989: p. 220; author's translation

mineral extraction, submarine naval technology and tourism appeared to offer good prospects. Their timing coincided with poor herring landings and the introduction of the 200 mile Exclusive Economic Zones which left Sweden's semi-industrial fisheries with reduced fishing opportunities in the Skagerrak and the North Sea. The likely impacts of this emerging situation on the coastal fisheries were assessed by the Commission in the 1980s. Their concerns in relation to coastal fisheries (see Box 16.1) remain largely relevant today and their recommendations were incorporated in an overall programme for Swedish Marine Resource Activities in the 1990s, submitted to the government in 1989 (Anon, 1989).

In the event, the industrial development of marine based activities envisaged in the 1960s and 1970s did not meet expectations. The reasons given for the demise of the Commission at the end of the 1980s differ according to those actively involved. One explanation was that cooperation between the relevant authorities was now sufficient and that legislation in the form of the Natural Resources Act provided the basic framework for the integration process. Others, however, suggest that sectoral rivalry and personal conflicts were undermining the Commission's work.

2.4 Comprehensive Municipal Planning

According to the *Planning and Building Act 1987*, each municipality is required to draw up a comprehensive plan for the use of land, water and the physical environment within its boundaries. The Comprehensive Municipal Plan is not a binding instrument but is expected to guide decision making in line with principles laid down in the Natural Resources Act and the Environmental Code and, in practice, to direct new development. Comprehensive Municipal Plans have a clear spatial development perspective and can be characterised as an instrument for dialogue between central government and the municipalities concerning the detailed implementation of the Natural Resources Act and which national interests are to take precedence in the development of the particular municipality. The County administration in fact represents central government interests and its key tasks are to scrutinise local plans to ensure that national interests are being respected and to focus on any transboundary issues.

The plans comprise maps and text describing the uses, regulations and recommendations. In theory they are intended to encompass territorial waters up to 12 nm from the baseline but in practice the seaward definition of the plans is often unclear and the criteria chosen for the outer limits vary considerably. While some plans cover the totality of the territorial waters, others refer only to the area within 3 miles of the baseline - in line with the remit of the Natural Resources Act - and yet others do not extend beyond the baseline itself.

The Swedish system of comprehensive municipal planning and its approach to coastal uses and to conflicts of interest has been analysed by Johansson (1995) and Ackefors and Grip (1995). In both instances the authors have chosen the plan prepared in the late 1980s for the Lysekil municipality on Sweden's west coast as their case study - although the representative nature of this plan can be disputed. The Lysekil plan is not typical: it belongs to a trial period of pilot projects coordinated by the Marine Resource Commission with access to special funds and expert support. Furthermore, at the time Lysekil was expected to play an important role in the development of the Swedish oil industry.

Comprehensive municipal planning in Sweden is commonly commended for the qualities of being decentralised, integrative in a vertical and horizontal sense, problem oriented and based on the democratic process and the long standing public review principle (Ackefors and Grip, 1995; Johansson, 1995). However, according to another assessment covering all coastal municipal plans (Boverket, 1995: p. 71) 'what happens or could happen on the coast is currently not prioritised in the municipal plans'. Another important reflection in this same assessment is that, in general, municipal plans mirror relatively few competing demands on coastal waters and there is therefore at the municipal level an impression that existing demands seldom result in conflictive situations. On the other hand, Boverket (1995) points out that competing claims over the use of coastal waters are often only indirectly evident in the text of the plans. In exemplifying this claim, 13 'hidden' conflicts were identified; of these 11 involved fisheries. In this context it may be perhaps more revealing to assess the significance not of the actual content of the plans, but rather what is in fact missing from the perspective of coastal fisheries.

As Ackefors and Grip (1995) argue, even though Sweden has a long standing tradition of the public review principle, planning remains a top-down process. This approach is at present under challenge from new forms of local participative experience. Among these new approaches are local Agenda 21 projects which involve citizens in environmental work, LEADER projects, a few in-depth Comprehensive Area Plans prepared at the local grass roots level, and consultation through focus groups. These initiatives are often led by individuals committed to the protection of the environment, the empowerment of local communities or simply to personal stakes in the plans' outcomes. They have also benefited local government at municipal and county level through a learning process in participative management.

In most instances, the Comprehensive Municipal Plan is prepared by officers of the municipality, with the support of a committee of locally elected councillors, experts brought in as consultants and specialised information made available at the provincial level or through research institutes. In general, architects tend to be overrepresented in the preparation of the plans most probably because this instrument is part of the Planning and Building Act and has its genesis in urban physical planning. This tradition also contributes to the rather static character of the plans, compared to process oriented concepts such as integrated coastal zone management. It also disconnects the planning process and the formulation of the plans - which are not in fact binding instruments - from the implementation process and the eventual outcomes.

In general, planners do not actively seek out the involvement of resource users. During the planning process, open consultative meetings are arranged and the community, in the widest sense, is invited to present its views, which also need to be submitted in written form. The attendance at public meetings is usually rather poor. Once drafted, the plans are exhibited in public for a period of two months during which time the planning process and its outcomes can again be questioned. Local fishermen are seldom active in the planning process: their needs are rarely separately assessed nor provide a stimulus for development proposals. Safeguarding their interests is usually left to the fishermen's organisations which will review the draft plans and bring to the municipality's attention whenever the plan appears to threaten essential fish habitats or where access to the fishing grounds could be disrupted by the plan proposals.

3. THE GOVERNING NEEDS OF COASTAL FISHERS IN RELATION TO COASTAL ZONE MANAGEMENT

The previous section has outlined the concern for the sustainability of the natural resource endowment which pervades Sweden's relatively sophisticated land use planning system. It has noted, in passing, that although this concern would seem central to the economic and social well being of Sweden's coastal fishers, they have rarely availed themselves of the opportunities to participate in the open planning process, preferring to leave the representation of their interests to the National Board of Fisheries or to their own professional organisations. The non-engagement of coastal fishers in local or national fisheries politics has been noted elsewhere and the reasons commonly cited include the difficulty of fitting attendance at public meetings into the abnormal working hours of the fishing industry; their distrust of - or even contempt for - bureaucracies which are considered the chief beneficiaries of the policy process; and self-perceived disadvantages when it comes to presenting their case in public debate.

But another explanation for the self-exclusion of fishers from the public policy process in relation to coastal zone management is rather more simple, namely that the issues considered central to coastal zone management and the planning system are at variance with the coastal fishers governing needs. According to McGlade *et al.* (1996) governing needs are the manifestations of problems experienced by individuals, groups or organisations. The starting point for the analysis of the coastal fishers' governing needs, is therefore, an inventory of these problems as perceived by the fishers themselves.

Preliminary studies indicate that localised, small scale fisheries along Sweden's west coast seem to be confronting two critical problems: first, how to increase the involvement of young people in the small scale fisheries - the problem of recruitment; and, second, how to legitimate their rights to fish in a multiple use coastal zone - the problem of resource access. The problem of recruitment manifests itself in a strong trend towards the increasing age of coastal fishers (Píriz, 2000) and is probably related to social processes, modern expectations of living standards and a tendency towards decreased regional dependency on local coastal fisheries. By contrast, the problem of access to sustainable fisheries refers to changes in the pattern of claims made on natural resources and is more diverse and dynamic. The results of a pilot investigation among 40 coastal fishers from the Swedish West Coast Fishermen's Association are summarised in Figure 16.2 which suggests that pollution, fisheries regulations and subtraction from the resource pool by other fishers - both professional and recreational - and by natural predators (seals) are perceived as the critical threats. Therefore a key question for

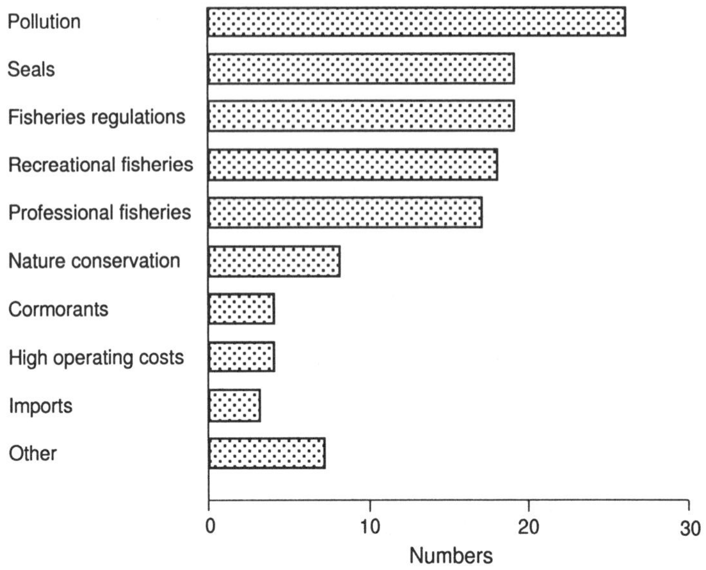

Figure 16.2 Perceived Threats to Coastal Fisheries

understanding the relationship between coastal fisheries and the coastal planning process is: to what extent are these threats articulated by the coastal management instruments discussed above?

3.1 The problem of youth recruitment

In rural coastal Sweden distances can be quite significant. Physical distance, combined with the changing social organisation of coastal fisheries (See Chapter 9), means that inter- and intra-generational communication within the coastal sector is becoming more difficult. With fewer inshore boats, the scaling down of crews to one or two men and the fact that catches are now commonly transferred directly from the quay to distant auction markets or processors, the local social networks have been truncated. One of the victims has been the transfers of knowledge from the older to the younger generation and the sharing of experiences among the younger generation. The lifestyle of the prototypical fisher which still survives on the islands around Göteborg, and which formally gained in value and prestige through the material, social and cognitive organisation of fishing practice within the community (Hazlehurst, 1994), has almost disappeared elsewhere. Coastal fishing has become a more isolated and less highly valued occupation.

With the boom in recreation and tourism combined with the unintended effects of the Shore Protection Act, land and housing along the coast has become scarce and the prices paid for coastal property well beyond the reach of young people. A highly competitive market has even been created for the small, shoreline cabins (workshops) traditionally used by coastal fishers to store and repair their boats and gear. For areas where summer houses currently account for 35% or more of local housing stock, central government has approved selective property tax rebates to assist local residents to acquire and maintain permanent homes in the coastal area. In Strömstad, the most northerly of Sweden's west coast municipalities, the revised plan recommends that priority be given in harbours to fishing rather than to sailing or tourism (Strömstad-Kommun, 1997).

3.2 Fisheries and marine pollution

Pollution is widely recognised as posing a serious threat not only to coastal fisheries (Figure 16.2) but also to recreation and tourism, and the problem is currently being addressed at all levels of governmental decision and through the Environmental Code's subordinated Acts and procedures. In general, eutrophication with blooms of opportunistic micro- and macro-algae and oxygen depletion is considered to be the principal pollution problem in coastal areas and the cause of structural and functional changes in shallow coastal ecosystems, affecting biodiversity and also lowering the economic returns from coastal fishing due to reductions in landings and higher operating costs resulting from clogged gears.

Despite all 'end of the pipe' efforts to reduce leaching of nitrogen and phosphorus from diffuse land based sources such as agriculture, problems still remain and fresh innovative approaches are needed. Between 1985 and 1995 discharges into the sea of nitrogen from agricultural sources were reduced by 25% (Regeringskansliet, 1998), half the national target. It is in this context that environmental management strategies and projects aimed at fixing and recycling nutrients from sea to land are becoming increasingly more important. Among such projects are the restoration of wetlands, shallow bays and fjords by removing built infrastructures which prevent the free circulation of water, harvesting large quantities of macro-algae and promoting mussel cultivation. However, what are proposed as solutions have a habit of resulting in the creation of new problems: the promotion of mussel cultivation is no exception (see Box 16.2).

Box 16.2 Looking for solutions but discovering new problems: the case of mussel culture

Constrained by low water temperatures, complex procedures for obtaining permits and strict environmental regulations, aquaculture development in Sweden has been slower than in other Nordic countries. Nonetheless attempts to fix and recycle nutrients from sea to land through the promotion of mussel culture confront a number of problems and objections. Since the algal blooms of the 1980s, harvesting of mussels has been subject to rigorous screening for potential risks to human health; a conflict has arisen over the methods of testing involving the incompatibility of EU and Swedish laws. Swedish law permits the use of biological tests only where there are no viable alternatives and although Swedish scientists claim that chemical analysis is at least as reliable as biological testing for toxins in mussels, this has yet to be accepted by EU countries which import the mussels.

Most mussel farms in Sweden use the longline technique of cultivation but the owners of summer houses along the coast have often objected to the location of mussel farms and influenced decisions on licensing. The real issue at stake here is the quality of the water for bathing, the aesthetic appeal of the coastal scenery and, above all, the market value of coastal real estate.

Recent applications to use seabed cultivation techniques have also run into opposition from coastal fishers - even though at the start of the twentieth century coastal fishers themselves collected and relaid mussels on small parcels of the seabed for use as bait. Today, however, questions concerning impacts on the marine ecosystem and on fishing are raised over plans to collect wild mussels from the Öresund area in the Kattegat and relay them on sites in the Skagerrak where favourable salinity conditions and good water circulation should enable them to grow larger and free of toxins.

Moreover the farming of shellfish means the granting of new exclusive rights which *de facto* extend beyond the allocation of seabed plots for the growing of mussels. According to the Fisheries Act, use of fishing gear is not permitted within 100 m of a fish farm. Thus any fish swimming in the vicinity of a privately licensed mussel bed fall, at least temporarily, under the protection of the licensee.

3.3 Coastal fisheries and the protection of marine wildlife

A recurring demand from Swedish fishermen is the reintroduction of 'protective hunting' (*Skyddsjakt*) to protect static fishing gear from predation by seals. Three different species are found in Swedish waters: the ringed seal (*Pusa hispida*), grey seal (*Halicoerus gripus*) and harbour seal (*Phoca vitulina*). Their populations have undergone drastic changes: they were abundant at the beginning of the century when hunting was permitted but, with industrialisation and the introduction of biocides (DDT and PCBs) and the effects of viral epidemics whose cause is not yet fully understood, seals became an endangered species and protection measures were adopted. General hunting was banned in 1974 though *skyddsjakt* was permitted until as late as 1987 when it was outlawed by the Helsinki Commission in the

Baltic Sea (Olsson, 1996). Recently populations have recovered and according to the fishing industry seals have again become a threat to viable fishing activities especially in coastal waters. Currently, the state compensates the industry for damage to gear and loss of catches to the tune of 6.3 million SEK per annum - a figure fixed some years ago and which no longer reflects the true costs.

Usually when it comes to the institutionalisation of nature's rights and the ascription of values to nature, central government plays the leading role, even though in most cases the initiative may have been triggered by international agreements or by pressure from established national or international NGOs. The issue of seals and coastal fisheries in Sweden has followed this pattern. In 1995 a working group, comprising the Swedish Environmental Protection Agency, the National Board of Fisheries, the National Museum of Natural History and the World Wide Fund for Nature, was established to consider the seal problem and seek solutions. Significantly, the Swedish Society for Nature Conservation - the leading pressure group - was unwilling to discuss any solutions which might involve restoring the hunting of seals.

The working group examined - and finally rejected - the hypothesis that a limited cull would have a significant effect on the scale of losses suffered by the fishing industry. From the perspective of the National Board of Fisheries the next logical step was to investigate ethically acceptable forms of *skyddsjakt*, though this was not universally supported by other members of the working group. Nor has the issue of seals and coastal fisheries been taken up at the level of the coastal municipalities. Indeed, in the perceptions of urban tourists seals are charismatic animals, and most coastal municipalities are today more dependent on tourism than on coastal fishing.

Similarly, the populations of cormorants (*Phalacrocorax carbo carbo* and *Phalacrocorax carbo sinensis*) have traditionally been kept in check by shooting and puncturing the eggs. Since joining the EU and, therefore, signing up to the Birds and Habitats Directives which give protected status to *P. carbo sinensis*, shooting and the destruction of eggs have been banned. However, in the region of Kalmar, south east Sweden, where the coastal fisheries are most affected by the cormorant populations, the County administration has recently decided to permit *skyddsjakt* in the vicinity of static fishing gears during the non-breeding season (August-March).

Another area of conflict between coastal fisheries and measures aimed at the conservation of marine wildlife concerns proposals by the Environmental Protection Agency to establish marine protected areas in line with policies endorsed by the Nordic Council of Ministers and the EU. So far protests by coastal fishers have focused on proposals to protect the Norwegian trench ecosystem with its high biological diversity in an area which coincides with a relatively stable shrimp fishery managed through locally agreed regulations. The imposition of an 'external' management regime risks the breakdown of the existing fisheries management system. On the west coast the County Administrative Board and the municipality of Strömstad have taken up the issue of a marine protected area and after negotiations with the local fishing industry have agreed to jointly develop a management system which both respects local fisheries management rules and protects Natura 2000 sites from the risk of damage by trawling.

4. CONCLUSIONS

It is clear that in Sweden several key planning instruments impact directly or indirectly on coastal fisheries in ways which may ultimately affect the coastal fishers ability to remain in business. Yet the highly respected model of integrated coastal management still tends to set aside the customary practices and informal rules developed by coastal fishers over the years.

What the foregoing analysis has indicated is that many of the 'threats' identified by the coastal fishers result from the failure to find common ground between different sectoral approaches to the concept of environmental conservation. In general, fisheries regulations are aimed at the conservation of fish stocks but pay little attention to other marine species or indeed to notions of social and spatial differentiation. Proposals for new forms of power generation are concerned with finding green alternatives in the energy sector but they pay scant attention to the disruption to ecological and socio-economic systems. The protection of marine wildlife species, involving a revalorization of natural resources, clearly has difficulty in accommodating the governing needs of other coastal zone users, including the coastal fishing industry. The failure to integrate the conservation policies of different user groups in general, and to raise the profile of coastal fishing interests in particular, poses some serious questions for the future of coastal fisheries and for the efficacy of coastal zone management in Sweden.

5. REFERENCES

Ackefors, H. and Grip, K. (1995) *The Swedish Model for Coastal Zone Management*, Swedish Environmental Protection Agency, Stockholm.

Anon (1989) *Svensk Havsresursverksamhet på 90-talet*, Delegationen för Samordning av Havsresursverksamheten, Stockholm.

Bernes, C. and Grundsten, C. (1991) *Sveriges National Atlas, Miljön*, SNA-förlag, Stockholm.

Boverket (1995) *ÖP-Analys Kust och Hav - en Utvärdering av Kustkommunernas Översiktsplaner*, Boverket, Planavdelning, Karlskrona.

Hazlehurst, B.L. (1994) *Fishing for Cognition: An Ethnography of Fishing Practice in a Community on the West Coast of Sweden. Cognitive Science and Anthropology*, University of Californa, San Diego, California.

Johansson, L. (1995) *Coastal Area Management in Sweden*, SWEDMAR, National Board of Fisheries, Göteborg.

Johansson, R. (1998) Personal communication, Senior Officer, National Board of Fisheries.

McGlade, J., Pauly, D. and Geronimo, S. (1996) *Interdisciplinary Scientific Methodologies for the Sustainable Use and Management of Coastal Resource System*, Final Report CEC Contract CTI CT 93-0349, Brussels.

Miljödepartementet (2000) *Levande Skärgård. Betänkande av Miljövårdsberedningen*. Statens Offentliga Utredningar: SOU:2000:67, Miljödepartementet (Ministry of Environment), Stockholm.

Olsson, M. (1996) *Säl och Fiske - en Gammal Konflikt i Östersjön*, Särgårdsinstitutet vid Åbo Akademi, Åbo.

Píriz, L. (2000) Dependence, modernisation and the coastal fisheries in Sweden, in D. Symes (ed.), *Fisheries Dependent Regions*, Blackwell Science, Oxford, pp. 116-131.

Regeringskansliet (1998) *Jordbruksdepartementet Pressmeddelande: Krafttag mot Övergödningen av Våra Vatten!* Pressmeddelande, Stockholm.

Riksdagen (1996) *På väg Mot ett Ekologiskt Hållbart Samhälle*, Riksdagen, Stockholm.

Segrell, B. (1995) *Den Attraktiva Kusten. Linköping*. Studies in Arts and Science, Theme Water, Linköping University, Linköping.

Svensk Författningssamling (1987) *Lag om Hushållning med Naturresurser m.m, SFS (1987:12)*, Norstedts Juridik, Stockholm.

Svensk Författningssamling Ds (1999) *Miljöbalken. SFS (1999:808)*, Norstedts Juridik, Stockholm.

Statistics Sweden (1998) *Statistical Year Book of Sweden 1998*, Statistics Sweden, Stockholm.

Strömstad-Kommun (1997) *Översiktsplan*, Stömstads Kommun, Strömstad.

Westerlund, S. (1992) *Miljökonsekvensbeskrivningar: Grunderna i Plan - & Marklagstiftningen, PBL, NRL, MKB*, Naturskyddsföreningen, Uppsala.

PART IV: CONCLUSION

Chapter 17

A FUTURE STRATEGY FOR INSHORE FISHERIES MANAGEMENT

Jeremy Phillipson and David Symes

1. INTRODUCTION

Systems of inshore fisheries management have tended to receive rather less critical attention within the wider critiques of fisheries management. As a result, there is a notable lack of material for comparative analysis or upon which to found effective policy decisions and development. This volume has taken a step towards redressing the balance, firstly by collating information in respect of seven EU Member States and, secondly, by highlighting a number of key issues common to most if not all the countries under study and central to the future direction of European inshore fisheries governance.

The inshore fisheries of the European Union display considerable complexity and diversity in terms of their biological, economic, social, cultural and institutional characteristics. Although commonly depicted as a marginal activity, the inshore sector does in fact represent a major component of the fishing industry - especially in terms of employment - and is conducted in a range of socio-political and geographical settings as diverse as the ice trap fisheries of the Gulf of Bothnia to the *étang* fisheries of the French Mediterranean coast. It is also highly differentiated. Thus, inshore fisheries encompass a range of activities including both capture and culture fisheries for both finfish and shellfish species. They may be undertaken on a full time, part time, seasonal or casual basis and they may be engaged in for commercial or recreational reasons. Inshore fisheries vary quite markedly between Member States and there are differing emphases between the relative importance of private fisheries, aquaculture and recreational fishing and in terms of the approach to management.

D. Symes and J. Phillipson (eds.), Inshore Fisheries Management, 297–310.

As the preceding chapters have amply demonstrated, the inshore fishing sector is structurally distinctive in economic and social terms. Predominantly comprising family based, small scale enterprises, it has a particular social organisation built on close social relations and a simple commodity mode of production, and is accompanied by specific forms of identity, modes of behaviour, divisions of labour and mechanisms for social reproduction which distinguish inshore fisheries from the more conventionally capitalistic mode of production for offshore fisheries.

The inshore fisheries sector may seem paradoxical in character. In some ways it displays characteristics that would infer resilience and flexibility; in others it would seem highly vulnerable and fragile. Thus, it has been argued that the sector commonly displays intrinsic strengths in terms of adaptive response, built upon a flexible system of social and work organisation, and therefore provides a vital contribution to the economic and social stability of peripheral communities. At the same time, however, the sector faces a large number of common challenges, including the relative immobility of the small scale boats, distance from markets, weak infrastructural provision, lack of social mobility, low levels of recruitment of young people and a marginal status within the industry in general.

Wider influences and external forces of change present challenges to the production system, traditional forms of adaptive response and processes of social reproduction within inshore fisheries. For example, the sector is not immune to the 'external' threats posed by global markets, resource depletion and regulatory strictures. Furthermore, it would seem particularly vulnerable to pressures for the privatisation and the concentration of fishing rights; this vulnerability is likely to be exacerbated by the peripheral location of many inshore fishing communities in national and regional space and the general trend towards the centralisation of economic activity. In less peripheral locations other 'externalities', derived from the multiple use of inshore waters, may place an added burden on local fisheries. The range of potential externalities is wide including pollution from the discharge of untreated urban and industrial waste waters, competition for space from other users of inshore waters and demands from conservationists for the enhanced protection of inshore ecosystems, habitats and wildlife species from the adverse impacts of fishing.

2. CONVERGENCE OR DIVERGENCE OF FISHING SECTORS?

In part, the lack of specific policy attention and comparative research into inshore fisheries reflects the simple truth that a common definition remains somewhat elusive. The conventions used in the Member States draw upon a range of structural, behavioural and spatial characteristics relating to the particular features of the inshore sector itself and to specific legislative, management and institutional traditions. Thus some definitions emphasise structural aspects - most commonly vessel length - but also vessel mobility, ownership characteristics and even levels of technical sophistication. Others focus on target species, the nature of the regulatory system, access rights or on elements of social organisation and culture. The 12 nm territorial limit would appear to provide a useful universal legal definition, in embracing most small scale fisheries and shellfish stocks, albeit that in Europe this zone is subdivided into two distinct areas in terms of access. Even so, the spatial definition clearly oversimplifies the nature and geography of the inshore fisheries. In general, the lack of a common structural definition and therefore of comparative data sets makes international comparison of the relative contribution of inshore fisheries extremely difficult. Nonetheless the national reviews contained in the present volume all serve to confirm the intrinsic importance of inshore fisheries in providing a significant share of the supplies of fresh fish and shellfish and, perhaps more fundamentally, in supporting both social and economic fabric within adjacent coastal regions.

Any attempt to outline a universal definition of inshore fisheries presents a complicated if not impossible task, especially when working within a cross-cultural framework. Nevertheless, what seems clear is that the challenge is one which is becoming more difficult with time. There are several forces which are impacting on the sector in such a way that it is arguably becoming increasingly difficult to identify as a discrete grouping within the industry. Five factors would seem significant.

- Firstly, technological developments affecting vessels, gears and fish finding equipment have not passed the inshore sector by; large parts of the sector are increasingly sophisticated, capitalised and mobile. Both spatially and structurally the sector is therefore becoming less easily distinguished from its 'offshore' counterpart. Inshore and offshore vessels will often fish side by side within and outside the traditional inshore waters.

- Secondly, convergence between inshore and offshore sectors would seem to be further intensified by the increasing incursion of offshore vessels into inshore fishing grounds given almost universal resource scarcity within the offshore finfish fisheries. In addition, there is evidence of owners of offshore fishing vessels diversifying their investments into inshore vessels, based on the perception of less restricted fishing opportunities in the inshore area.

- Thirdly, in regulatory terms a key consideration is whether management systems themselves encourage or discourage the distinctive features of the inshore sector. It has been suggested elsewhere (Salmi *et al.*, 2000), for example, that the part time, seasonal or polyvalent nature of many inshore fisheries is becoming increasingly marginalised within modern management systems, by the tendencies towards specialisation, by over-simplified and blanket regulatory approaches and for reasons of convenience in administration or effectiveness of monitoring and enforcement. This presents a challenge to policy makers in developing approaches to inshore fisheries governance which can facilitate rather than subjugate the traditional flexible strategies of the inshore fishing enterprise.

- Fourthly, it would seem that inshore fishing communities themselves - their structures, social organisation and culture - are in a state of flux resulting from a range of social processes incorporating new aspirations, living standards and lifestyles within the fishing household. The geography of demographic change within fishing communities is also differentiated, for example between those in rural peripheries facing depopulation and those exposed to the infiltration of new urban inhabitants - including retirees and commuters. For the latter, not only is the inshore fishing community increasingly difficult to distinguish from other fishing interests, the fishing community itself is often increasingly invisible within the wider coastal community (Nuttall and Burnett, 1998).

- Finally, it may also be important to question whether the social organisation of the inshore fisheries sector is itself being reproduced in new forms, which may call for a refocusing of analysis. For example, in some Member States there would seem to be little evidence remaining of the traditionally widespread dual economy of farming and fishing which has been a defining characteristic of many peripheral rural economies. To what extent, however, has the pluriactivity of such economies been reinvented within new forms of organisation, embracing sectors such as tourism? Here there is a pressing need to reassess the concept of a pluriactive coastal economy and to examine the implications for inshore fisheries of new forms of work combination within the fishing household.

There seems little doubt that the inshore fisheries sector has been influenced by the processes of modernisation, globalisation and social change, and that this is making the boundary between inshore and offshore fisheries less clear. Nevertheless it also behoves the social scientist to test the alternative hypothesis that, despite the blurring of the edges of conventional definitions, inshore fisheries remain fundamentally different from offshore fisheries and should therefore be accorded distinctive modes of management which maximise the social utility of inshore fisheries while protecting the highly valued and sensitive coastal ecosystems in which they are situated. Circumstantial evidence points to the existence throughout Europe of large numbers of small scale, labour intensive, family based fishing enterprises. But their distinctiveness is not simply a matter of scale. If Hansen and Højrup (Chapter 11) are to be believed, the whole ethos of small scale, coastal fishing is diametrically opposed to the conventions of the capitalist firm in its approach to production, investment and efficiency. Moreover, the economic significance of the inshore sector is not to be measured in terms of the gross value of landings or the total employment generated, but rather in its contribution to local economies often in deprived, remote rural areas. Seen in this light, the fact that 80% of the EU fleet is under 12 m - and more than half is under 9 m - it is not so much a symptom of unfulfilled economic rationalisation nor an imperfect division of labour, but a testament to the importance, resilience and relevance of inshore fisheries.

3. THE COMPLEXITY OF INSHORE FISHERIES MANAGEMENT

Developing effective approaches to the management of inshore fisheries represents a complicated task. At the detailed level of policy implementation, monitoring and enforcement, the difficulties are particularly apparent given the fragmented nature of production, the large numbers of often unregistered production units, their fluctuating activity rates and their often low and uncertain catch quantities. More fundamentally, inshore fisheries are host to the whole range of social, economic, political and biological challenges which beset fisheries management systems generally; most of these challenges are intensified in an inshore context and are arguably much more complex.

In part, the added complexity of inshore fisheries management arises from the overlapping nature of the governance regime whereby local, national and European levels of regulation intersect. In the EU the inshore domain has become the locus for the delicate interplay of supranational and nation state competencies (embodied in the principles of proportionality and subsidiarity) and in some instances the setting for experiments in local governance and user participation. As a result in most Member States the statutory basis for inshore fisheries management is divided among a number of responsible organisations and across a range of spatial scales. Effective coordination of these different organisations and scales represents a major challenge for fisheries managers.

In practice, the current access derogation in respect of the 12 nm limits implies considerable discretion for Member States to determine their own approaches to inshore fisheries management. In effect, management and organisational structures for inshore fisheries reflect the particular institutional and organisational traditions within the Member States and the specific structural and behavioural characteristics of their sectors. The resulting regulatory regimes utilise complex systems of property rights and both statutory and informal means of regulation. Some states have adopted decentralised and delegated approaches which allow for the involvement of local user groups. Others rely largely on a state led centralised management approach regulating the inshore sector on the basis of fleet segments and gear groups. However strongly modern theories of governance may argue in favour of devolved, participative management, there is no clear evidence from the national studies that state led, centralised management has in any way seriously disadvantaged the inshore sector. While inshore fishermen in those countries which have built up a tradition of locally devolved management would not wish to see this privilege rescinded, in other countries there is no insistent demand for decentralised or devolved management.

The complexity of inshore fisheries governance is not solely related to questions concerning the distribution and coordination of governing responsibilities. In addition, there are dilemmas relating to multiple stakeholder interests in the coastal zone. On the one hand, this poses significant challenges to establishing effective systems of user participation at the local level, with issues being raised over which user groups have a legitimate stake in fisheries management. On the other hand, and more fundamentally, there are additional objectives and issues which enter the fisheries management equation in an inshore fisheries context. For example, inshore fisheries are clearly at the forefront of the burgeoning interest in environmental and ecosystem management (Symes and Phillipson, 1997; Coffey and Dwyer, 2000). In addition, the contribution of inshore fisheries to the economic and social stability of many coastal communities, means they have a particularly important role in issues of regional dependency and the development of remoter coastal regions (Symes, 2000). Inshore fisheries are therefore the subject of differing claims as to their roles and functions, which go far beyond the basic purpose of food production. Resolving these issues demands a careful integration and balancing of objectives and approaches between fisheries and other policy domains.

Within the last decade or so there has been a surge of new ideas concerning the future development of fisheries management. Among these are concepts of rights based management (see, for example, FAO, 2000), regionalisation (Symes, 1998a), an ecosystem based approach (Grumbine, 1994; NMFS, 1999) and co-management (Jentoft, 1989; Sen and Raakjaer Nielsen, 1996). Most have important implications for inshore fisheries. With the exception of rights based management which at first glance appears to reassert the 'dominant use' approach to managing the living resources of the sea, each of these new concepts, to a greater or lesser degree, embraces the quest for a more integrated system of fisheries management. Yet until quite recently fisheries managers have been slow to re-evaluate the role of inshore fisheries, reluctant to embrace environmental or social objectives and overly cautious about adopting new institutional frameworks for management.

4. REASSESSING INSHORE FISHERIES MANAGEMENT

In our view the nature of inshore fisheries - and their importance in economic, social, cultural and environmental terms - calls for a distinctive and highly focused approach to their governance. The original derogation from the principle of equal access in respect of the 6 and 12 nm limits was predicated on the specific characteristics of inshore fisheries, their contribution to the local coastal economy and the need to protect the resource base from unwarranted exploitation by the more mobile offshore fleets. As the Commission's Green Paper on *The Future of the Common Fisheries Policy* (European Commission, 2001) makes clear, those circumstances are as relevant today as they were almost twenty years ago. Accordingly their recommendations are for maintaining the *status quo* in terms of access agreements and strengthening the coastal state's role in relation to the management of the 0-12 nm zone.

Any alternative move would have widespread ramifications for the future of Europe's inshore fisheries. Indeed, the first of the alternative scenarios - the removal of the derogation - would represent a major transformation in the political and social makeup of fisheries governance in the EU. Although in accordance with the prior principle of equal access throughout EU waters, such a move would be strenuously opposed by a majority of Member States as a major breach of well established lines of national interest and the abrogation of the sense of 'ownership' of the management of inshore waters. Moreover, it would undermine the undoubted efforts on the part of some Member States both to conserve stocks in inshore waters and to develop systems of local equity by exposing inshore fishing communities to the full impact of open access. A second scenario, which does not seem to have attracted much political support, would be for the renewal of the access restrictions on a permanent rather than temporary basis. The potential benefits of such a move would seem strong. A permanent derogation would further encourage the development of dedicated management regimes for inshore fisheries by Member States and allow a greater degree of long term planning for both fisheries managers and the fishing industry alike. Indeed, there are those who would argue that failure to cement the current access agreements into a permanent state of affairs is likely to consign inshore fisheries management not simply to disabling uncertainty but to the very real risk of ending the coastal state's independent management role within the next ten to twenty years. The significance of this possible train of events is thrown into sharp relief by comparison with current practices within the Common Fisheries Policy (CFP) where policy measures have generally been applied across the board for both inshore and offshore fisheries. There would be very little confidence among inshore fishermen anywhere in Europe that centralisation of inshore management at the EU level would permit the sensitive handling of what are commonly local management issues.

While some may argue for a further extension of the EU's role in inshore fisheries management - through the setting of broad objectives and developing a common framework for Member States' roles and responsibilities - in most cases inshore fishermen would resent any further intrusion by the European Commission. No one could claim that the coastal states have made an unqualified success of inshore fisheries management under the existing conditions but there is sufficient evidence to suggest that they have not made the serious errors of judgement commonly laid at the door of European institutions. Perhaps the one area where greater intervention by the EU would be welcomed is the provision of more clearly focused financial support for inshore fisheries initiatives. Appropriate measures might include support for the adoption of more selective and ecosystem sensitive fishing gears in inshore waters, the funding of local management schemes aimed at the conservation of commercial stocks or the enhanced protection of marine wildlife habitats and species, and the provision of adaptation grants to compensate inshore fishermen for the loss of income while adjusting to new conditions imposed, for example, by the introduction of 'no take zones'.

The Commission's Green Paper indicates a willingness to recognise inshore fisheries as a special case both in terms of structural policy and in its recommendation to extend the coastal state's powers to adopt stricter conservation measures through the 12 nm zone. But it reserves its strongest endorsement of the distinctive identity of the inshore sector in relation to financial measures aimed at reducing the capacity of Europe's fishing fleets. Here, it argues that

> 'Because of their importance for employment, in particular in local areas with few alternative opportunities, and because they have, if properly managed, a lower impact on the resources, small-scale artisanal fisheries may need to be excluded from this general approach. Such fisheries could be the beneficiaries of a specific aid programme, subject to clear conditions for eligibility, including common definitions of "small-scale" fishing activity and "fisheries-dependency" of a coastal zone, and limited impact on competition between the Member States' fleets.'
>
> (European Commission, 2001: pp. 32-33)

4.1 A zonal management approach to inshore fisheries

Despite this endorsement of inshore fisheries - or more accurately the small boat sector - as worthy of separate consideration, the European Commission has largely been caught up in the general indecision over the future of inshore management. Rather than simply making *ad hoc* adjustments to existing policies there is an argument to be made for a much more focused approach to the management of inshore fisheries, separate at least in detail from that applied to the inshore sector. The idea of a zonal management approach to EU fisheries, whereby inshore and offshore management strategies are developed along different lines, underpinned by different priorities and guided by different combinations of policy measures has been posited earlier (Symes, 1998b). Figure 17.1, adapted from the earlier publication, describes the main features of zonal management. The overarching framework would be provided by a regional fisheries management strategy and informed by an ecosystem based approach. In the interests of integrated fisheries development at the local level - and also for marine environmental considerations - it is important to incorporate aquaculture within the zonal approach, notwithstanding the facts that it is organisationally distinct from other forms of coastal fisheries and will certainly be subject to very different forms of regulation. As Figure 17.1 indicates, under these proposals inshore capture fisheries would require a combination of restrictive licences and community based quotas, supplemented where appropriate by individual use rights for the harvesting of shellfish and subject to the direction of a responsible local organisation. The guiding principles for management would be to maximise the social utility of the inshore fisheries for the local community (employment and income), while guaranteeing the diversity, productivity and integrity of the inshore ecosystems. In rather marked contrast, offshore management would be centrally directed by the relevant regional authority and could pursue a

	Inshore Fisheries (0-12 nm)			Offshore Fisheries (12-200 nm)
	Shellfish	Finfish	Aquaculture	
Biological parameters	Sessile (bivalves); low mobility (crustacea)	Highly mobile		Highly mobile
Socio-economic parameters	Artisanal	Artisanal	Artisanal/industrial	Industrial
Use rights	Individual/community • preferential licensing • several rights	Community based • preferential licensing • group quotas	Individual/private • licensing	Individual • ITQs
Regulatory systems	Local regulation • licences (conditions attached) • gear regulations • minimum landing size (MLS) • closed seasons/areas	Local regulation • vessel access (size) • gear regulations • ground closures Centralised regulation • quota management • MLS/bycatch	National regulation • environmental standards Local regulation • designated areas	Centralised regulation • TACs and quotas • gear regulations • MLS • bycatch regulations • ground closures
Management authority	Local institutions	Local/national/EU regions	National	EU regions
External influences	• regional economic development programmes • integration with marine environmental management or CZM			• international treaties • NEAFC
		• EU regional fisheries strategies	• water quality objectives	

Figure 17.1 A Zonal Management System for EU Fisheries

strategy of economic rationalisation based on the allocation of individual property rights. Again, it would be necessary to put in place some safeguards for the sustainability of the wider marine ecosystem.

Zonal management raises a number of issues concerning its implementation and the distribution of competencies between the EU, individual Member States and local institutions, particularly in the management of finfish fisheries within inshore waters. At the very outset of this volume it was made clear that inshore fisheries management cannot be completely divorced from the attempts to regulate offshore fisheries. A 'de-coupled' management system would require close integration of all levels of intervention to ensure overall coherence. If such a system were to involve separate institutional structures including local management organisations, continuing communication and coordination between different parts of the system would be essential.

Proposals to extend coastal state regulation out to the 12 nm limits raise some intriguing questions concerning the impact on inshore fisheries and their management. From a resource management perspective it makes good sense; it also reflects the extended range of inshore fishing activity. But the very presence of non-coastal state vessels within the 6-12 nm zone serves to contradict the notion that these are local waters appropriately regulated by local management institutions. In this outer zone there is a much stronger emphasis on finfish exploitation and the deployment of mobile gears by nomadic fleets of larger vessels. Such considerations must raise some doubts about the justification of transferring *de jure* responsibility to local management institutions. Yet the alternative case for dividing responsibility between local and central institutions has little to commend it. Paradoxically the proposed development could weaken rather than strengthen the concept of locally based management for inshore waters.

4.2 Integrated fisheries management

At several points in this book a plea has been made for more integrated management. But what precisely does this term mean? It has at least two quite distinct interpretations. On the one hand it refers to the need for a more holistic approach to fisheries management in which the various biological, economic, social, cultural and ecological objectives are brought together in a more balanced and inclusive way. The argument rests on the premise that to date fisheries management has been driven very largely by a scientifically rational despotism, combining the imperfections of fisheries science and economics in somewhat unequal measures, to the detriment of socio-cultural and ecological goals and values (Symes, 2001). In the context of inshore fisheries, a more inclusive assessment of the needs of the fishery could lead to the prioritisation of social and cultural objectives over the more familiar goal of economic efficiency. On the other hand, integrated management is also linked to the need for closer collaboration and coordination with other policy sectors concerned with the management of marine space and its natural resources. Again the idea has a special resonance for inshore waters where there is greater competition for limited space and where conflicts of interest are more sharply focused (see Chapter 15); here it is intimately associated with notions of integrated coastal zone management.

Each of these concepts embodies an ambitious project, but neither is particularly new, thus suggesting that while they may be seductively persuasive on paper they remain stubbornly resistant to consummation in practice. The same fate could well befall another neologism - the ecosystem based approach to fisheries management (see Chapter 14) - not least because the concept is as yet only partly formed and thus subject to a very wide range of interpretations (Yaffee, 1999). Despite its infancy it is already quite firmly associated in the minds of the fishing industry with greater intervention by the state, more regulation and still further restrictions on the rights to fish. These attitudes will be difficult to overcome. In the longer term the catalyst for integrated fisheries management may lie in a radically new approach to the science of fisheries management. The transformation has already begun with the incorporation of the precautionary approach and, notwithstanding the considerable technical problems involved, an increased level of interest in multi-species modelling and assessment. Such developments lie at the core of an ecosystem based approach though they will need to be extended to include non-commercial species. Progress towards a more fully comprehensive system of integrated management is, however, bound to be slow, uneven and, for some, a highly charged and frustrating process.

It may eventually prove possible to develop a system of integrated management based on an ecosystem approach which sympathetically addresses issues of marine wildlife conservation, but for the present it is difficult to visualise just how such a system will be institutionalised. Not surprisingly, inshore fishermen are bemused by and suspicious of the terms 'integrated management' and 'ecosystem based approach' and therefore sceptical when it comes to claims that they will necessarily enhance the sustainability of commercial fisheries in the longer term. Most would probably welcome more dialogue with other interest groups so as to better understand each other's positions and in the hope of finding firmer common ground. But they are certain to resist attempts to develop new institutional structures for the governance of marine resources for fear that this would further diminish the identity of the inshore fishing industry and reduce the attention paid to fisheries management *per se*.

Thus we end this review of inshore fisheries management in Europe, more or less where it began, with the issue of the identity of inshore fisheries - but on this occasion with concern for its political profile. In most European countries inshore fisheries lack a strong, independent voice: they struggle to gain recognition beyond the rather narrow, parochial confines of local politics. National federations of fishermen's organisations are arguably dominated by the large boat, offshore industry whose interests rarely coincide and may indeed conflict with those of the small boat sector. There is therefore a danger that when refracted through the opinions of such organisations, inshore interests may become distorted.

To some extent the political weakness of the inshore sector is an inherent condition of its individualistic, small scale, family based structure, for where the fishing enterprise is constructed around the dwindling human resources of the family there will be little spare capacity for engagement even in local - and still less in regional or national - associational activity. In some countries, it is the fishermen's wives who have added to their range of fishing related activities by becoming involved in lobbying on behalf of the inshore industry. Solutions to the question of raising the political profile of inshore fisheries are elusive. Help could be forthcoming from a somewhat unexpected source, namely some of the high profile environmental interest groups who perceive the small boat sector as offering the best prospects for developing an environmentally responsible approach to fishing.

5. REFERENCES

Coffey, C. and Dwyer, J. (2000) *Managing EC Inshore Fisheries: Time for Change*, RSPB, Sandy.

European Commission (2001) *The Future of the Common Fisheries Policy*, Commission of the European Communities, Brussels.

Food and Agriculture Organisation (FAO) (2000) Use of Property Rights in Fisheries Management, *FAO Fisheries Technical Paper* 404/1, FAO, Rome.

Grumbine, R.E. (1994) What is ecosystem management? *Conservation Biology*, **8** (1), 27-38.

Jentoft, S. (1989) Fisheries co-management: delegating government responsibility to fishermen's organisations, *Marine Policy*, **13** (2), 137-154.

NMFS (National Marine Fisheries Service) (1999) *Ecosystem Based Fisheries Management*, A Report to Congress by the Ecosystem Principles Advisory Board, Department of Commerce, Washington.

Nuttall, M. and Burnett, K. (1998) The negotiation and management of crisis in the Scottish fishing industry, in D. Symes (ed.), *Northern Waters: Management Issues and Practice*, pp. 238-246.

Salmi, P., Salmi, J. and Lappalainen, A. (2000) Finnish part-time fishery: a flexible local strategy or a nuisance for fisheries management, in D. Symes (ed.), *Fisheries Dependent Regions*, Blackwell Science, Oxford, pp. 132-142.

Sen, S. and Raakjaer Nielsen, J. (1996) Fisheries co-management: a comparative analysis, *Marine Policy*, **20** (5), 405-418.

Symes, D. (1998a) Towards 2002: subsidiarity and the regionalisation of the Common Fisheries Policy, in T.S. Gray (ed.), *The Politics of Fishing*, Macmillan, London, pp.176-193.

Symes, D. (1998b) Towards a property rights framework for the management of Europe's fisheries, in D. Symes (ed.), *Property Rights and Regulatory Systems in Fisheries*, Blackwell Science, Oxford, pp. 257-264.

Symes, D. (ed.) (2000) *Fisheries Dependent Regions*, Blackwell Science, Oxford.

Symes, D. (2001) *Integrated Fisheries Management: A Challenge for the Common Fisheries Policy, The Buckland Lectures, 2000*, Buckland Foundation, London.

Symes, D. and Phillipson, J. (1997) Inshore fisheries management in the UK: Sea Fisheries Committees and the challenge of marine environmental management, *Marine Policy* **21** (3), 207-224.

Yaffee, S.L. (1999) Three faces of ecosystem management, *Conservation Biology*, **13** (4), 713-725.

INDEX